THIRTY-THREE
WAYS OF
LOOKING AT AN
ELEPHANT

THIRTY-THREE
WAYS OF
LOOKING AT AN
ELEPHANT

EDITED BY **Dale Peterson**

Trinity University Press | *San Antonio*

Trinity University Press
San Antonio, Texas 78212

Book design by BookMatters
Cover design by Rebecca Lown
Cover image: MF012 / Shutterstock
Half title page image: Donvanstaden / iStock
Title page image: Donovan van Staden / Alamy Stock Photo
Last page image: Evgeny555 / iStock

Printed in Canada

ISBN 978-1-59534-866-1 paperback
ISBN 978-1-59534-867-8 ebook

Trinity University Press strives to produce its books using
methods and materials in an environmentally sensitive manner.
We favor working with manufacturers that practice sustainable ·
management of all natural resources, produce paper using
recycled stock, and manage forests with the best possible
practices for people, biodiversity, and sustainability. The press
is a member of the Green Press Initiative, a nonprofit program
dedicated to supporting publishers in their efforts to reduce
their impacts on endangered forests, climate change, and forest-
dependent communities.

The paper used in this publication meets the minimum
requirements of the American National Standard for
Information Sciences—Permanence of Paper for Printed Library
Materials, ANSI 39.48-1992.

CIP data on file at the Library of Congress

24   23   22   21   20 | 5   4   3   2   1

To Iain Douglas-Hamilton, Cynthia Moss, Joyce Poole,
Raman Sukumar, Andrea Turkalo, and all other heroes
for their work on behalf of elephants

CONTENTS

FICTIONAL AND LITERARY ELEPHANTS

PREFACE

For more than three decades, I have written about animals and con-servation. A few years ago, it happened that a major publisher was planning to produce a large-format book of beautiful elephant pho-tographs along with an informative background text. I was asked to write the text.

Thus began my education in elephants, which was among the most transformative experiences of my life. First, I sought the education of direct experience: to see, hear, smell, touch, and if possible commu-nicate with elephants. The book's photographer, Karl Ammann, and I traveled to Asia and across Africa in order to come into direct contact with individuals of all three species: Asian, African savanna, and African forest elephants.

We met Asian elephants by inserting ourselves into a teak-logging operation, in the northwestern mountains of Myanmar, that still used trained elephants to do things in places expensive logging machines cannot go. We rode elephants for two days out to the logging camp, and then we rode elephants for two days in return, so I got to know them up close, discovering, for example, that an elephant's tongue is about the size and texture of a well-padded softball glove. I found that elephants are approximately as surefooted as mountain goats, which explains a good deal about how Hannibal crossed the Alps with war elephants in 218 BCE. I saw that when an elephant slips on one foot, there are still three left over to recover with. I learned that nothing in the world takes precedence over a baby elephant's needs.

In Africa, the photographer and I followed and took pictures of savanna elephants in the magnificent savannas of Kenya (at Samburu National Reserve and Amboseli National Park) and of forest elephants in the swampy *baï* (clearing) of Dzanga-Sangha National Park in the Central African Republic.

The three species (*Elephas maximas* in Asia, *Loxodonta africana* and *Loxodonta cyclotis* in Africa) are different enough in outward appearance that a nonexpert can soon learn to tell them apart, but individuals of all three are equally astonishing to see in real life. They have some almost magical qualities. They seem to exist behind their noses and ears, rather than behind their eyes as we people do, and they can move slowly enough to test the patience of most ordinary human beings—or run fast enough to challenge anyone's fortitude. They can generate a deafening blast of sound, when the occasion calls for it, or they can be as quiet as a soft breeze for as long as they choose. They can appear when you least expect them, and they may disappear when you don't. They live in families with angelic babies, wildly playful youngsters, and fiercely protective moms, all guided by the sober leadership of grand matriarchs.

Among elephants' most obvious features are their formidable size and strength. Their amazing intellectual, cognitive, and emotional qualities are less obvious and yet, I think, far more interesting and worth knowing about. And they are best learned not from direct contact but rather from indirect contact, through examining the writings of scholars, scientists, thinkers, and doers with elephant experience. Such an examination was, of course, the second part of my education in elephants, and it is what I recapitulate in this reader, which is, I believe, the only significant anthology of primary historical, cultural, and scientific writings about elephants.

I have included folktales from across Africa, sacred commentary from ancient India, historical recollections from Burma (now Myanmar), informed fiction from mid-twentieth-century Vietnam, children's literature from post–World War II Japan, classical writings that originated in Greece and Rome, a piece translated from Latin by medieval scribes, exposés of abuse written by Indian and American journalists, colonial writings fresh from the pens of ferocious ivory hunters, and several samples of the scientific literature from the second half of the twentieth century. I have chosen all these selections based on literary and dramatic qualities as well their relevance in the long history of elephants, although it is, to be more accurate, the long history of elephants as people have seen and thought about them. The true history of elephants—divorced as it must be from the bias and ignorance, greed and cruelty, hopes and dreams that so characterize our own species—has yet to be told.

CULTURAL
AND CLASSICAL
ELEPHANTS

The Meaning of Elephants

DONALD J. COSENTINO, JOYCE POOLE,
COLIN M. TURNBULL, AND JAN KNAPPERT

Elephants and their ancestors, all members of the order Probosicidea— the proboscideans—emerged from the mammalian line some 60 million years ago in the form of such exotic creatures as the *Moeritherium*—a Latinized coinage honoring "the Wild Animal of Moeris." Moeris was the name of an ancient lake at the bottom of a vast sinkhole in Egypt where, at the start of the twentieth century, the British paleontologist Charles Andrews found fossil remnants of that particular beast. The Wild Animal of Moeris was about the size of a pig and the shape of a hippo, with eyes and ears situated high in an elongated skull that ended with four front incisors extended forward into four small tusks. The tusks guarded, it seems, an extended and flexible snout: a proboscis. Other comparable fossils more recently discovered in northern Africa suggest that early proboscideans may have wallowed in warm and shallow waters: the opening actors in an extended, successful, and remarkably diverse evolutionary drama that produced 8 families, 38 genera, and more than 160 species of trunked and tusked creatures who became dominating inhabitants on every continent except Antarctica and Australia.[1]

Not long after modern humans learned to make penetrating stone tools and developed the skills of group hunting, a wave of large-animal extinctions swept the planet, wiping out half the genera of mammals weighing more than 40 kilograms, including all but a few of the dozens of large terrestrial proboscideans then alive. The extinction of a remnant group of mammoths surviving on the Wrangle Islands off the coast of Siberia, completed around four thousand years ago, left only three final species. Those three are Asian elephants (*Elephas maximas*) and the African savanna (*Loxodonta africana*) and forest (*Loxodonta cyclotis*) elephants.[2]

Elephants are still alive, and with some effort you and I can still see and even, under special circumstances, touch them. We can experience them in the waking reality of the present. But they are going fast. For most people in most parts of the world, they are already gone. They are no longer present in the wild or part of the reality of people's daily lives, represented today pathetically by a few isolated prisoners in zoos and circuses, remembered dreamily in a debased iconography as winsome Dumbos and Jumbos. In those parts of the world where wild elephants are still alive, in scattered patches of Africa and Asia, they are encountered as real, actual creatures who provoke fear and distress as well as awe.

When we wonder about the *meaning* of elephants, we are asking a human-centered question: What is their meaning to us? It is common, of course, to speak of their meaning to us in crudely pragmatic terms—as a source of meat, ivory, wealth, power, and so on. But they can also mean something deeper, more complex and more elusive. As the grandest terrestrial animals on the planet, they may serve as icons or symbols or types.

We might imagine them as kings of the forest, mighty representatives of authority, or the dark embodiments of danger and death. They dominate the landscape with their size, power, and destructive capacity; and, in Africa, even in those many parts of Africa where they once were and no longer are, their echoes remain, culturally expressed in sculpture and painting and carving, in masks and masquerades and songs—and perhaps above all in the narrative arts: folk tales. This chapter provides a transcontinental sampling of contemporary and traditional elephant folk tales from Africa.

The opening piece of that sampling is a fragment from a longer Chadean tale that describes how a hunter once spied on several young women bathing in a river. After they emerged from the river, the hunter saw, the women slipped into elephant skins and, thus transformed back into their original elephant selves, ran off to join the rest of the herd. The hunter returned the next day to hide the skin of the most beautiful of the bathing women. Unable to return to her elephant self she was left behind, a mere woman, whereupon the hunter revealed himself and married her. Years later, after she had discovered her old skin and thus understood her husband's trick, she took her revenge.

"How Elephants Came to Eat Trees" is a Samburu story, while the following one, "The Bride Who Became an Elephant," was told by a member of the Maasai tribe. The Samburu and the Maasai are both East African pastoralists, and both maintain a cultural taboo against eating elephant meat. They sometimes speak of recognizing a special affinity between people and elephants based partly on the logic of physical and psychological continuity.

"The Pygmies and the Elephant" and "Why Elephants and People Can Never Be Friends" come from the Mbuti people (or BaMbuti) of the Ituri forest in the eastern Democratic Republic of the Congo (DRC). The BaMbuti are part of a larger association of people living in the Congo Basin who are commonly known as Pygmies. Pygmies are forest specialists, skilled in extracting the necessities of their daily lives from the forest by hunting and gathering; the BaMbuti are particularly known as net hunters.

Net hunting for the BaMbuti—driving game into long nets—takes place on most days of the week. On the days when they are not hunting, the men repair the nets, make string for the nets from vine fibers, and relax. Women gather vegetables and cook on every day of the week. But there is also the honey-gathering season, a time when everything inessential is forgone and when BaMbuti camps in the forest are, according to anthropologist Colin Turnbull, "filled with singing and dancing day and night."[3] This is also the time for telling stories. The Pygmies live in a social world frequently burdened by precarious relationships with neighboring Bantu tribes, people who are gardeners rather than hunters and gatherers and who may fear the forest and despise their physically smaller neighbors, the forest people. In the tale "The Pygmies and the Elephant," this social tension is one aspect of the problem with the MuBira man (a member of the neighboring Bantu tribe known as the BaBira), who turns into an elephant in order to attack his wife.

Finally, "The Origin of Mankind," a long narrative from which I have presented a brief fragment, comes from the Ngbandi people living in Equateur Province of the northwestern DRC. It is worth noting that one of the complexities of this tale derives from multiple puns potentially embedded in the name Tondo-lindo. Ngbandi is a tonal language, which means that the word *tondo*, referring at first to a "nicely ripe and red" fruit, can be broken apart syllabically and played with tonally to indicate "father" (*to*), "forest" (*ndo*), "testicle" (*toli*), "clitoris" (*linda*), "seed" or "fruit" (*li*), or the verb "to enter" (*li*).[4]

All narratives express the dance between dream and waking. Even the most prosaic of stories, presented as an honest and rational recounting of some real event in the waking world, represents in fact one person's fleeting impressions passed through the baffle of perspective, memory, and language. A story may be true and have meaning, but how true and which meaning? Stories repeated become tales. Tales dramatized beget fables and myths. And in Africa, where spoken narrative is an important art form, the words, the very words here pressed and desiccated on the page, will be expanded by gesture and mime, enlivened with song, and invigorated through repetition and spontaneous improvisation. "Domei o domeista!" (Story, stories!), a Mende audience in Sierra Leone shouts gleefully, and the storyteller steps forth to perform.[5] Nothing is perfectly scripted, nothing told the same way twice, and why should it be otherwise?

The Day the Elephant Wife Took Her Revenge

That day she was so excited she worked quickly, quickly. This surprised many people who asked her if she was going somewhere. She just laughed. Then she plaited her hair, took her bath, and oiled her body with vegetable oil. Next she bathed her children. After doing this, she took her skin and put it on. She took a pestle and threw it: *peenre!* And the pestle became her trunk. She let fall the mortar: *dii!* She took another and let it fall: *dii!* And her four feet appeared. The children ran in all directions crying: "Our mother has become an elephant! Our mother has become an elephant!"

She broke the straw fence, and started running with heavy steps and disappeared into the bush. She breathed in the forest air with furor and found her husband. She caught him, tore him, and threw him away.

How Elephants Came to Eat Trees

Long ago, Elephant said to God, "I am bigger than you, and I will eat animals." God replied, "No, you will eat trees," to which Elephant refused.

So God said, "Can you beat me?" and Elephant replied, "Yes, if you rain, I can rain."

And God said, "Let me see you rain. I want to see the water run."

So Elephant made a small hole into which he urinated. When he finished, God asked, "Have you rained?" to which Elephant replied, "Yes."

Then God said, "Make a light as a signal," and Elephant wagged his head so that his tusks moved back and forth, and God asked, "Have you made a signal?" to which Elephant replied, "Yes."

And God said to Elephant, "Can you shout?" and Elephant made a mighty sound.

Then God said, "Let me hear you shake in your stomach," and Elephant made a rumbling noise in his stomach.

Finally God said, "Wait for me!" And He brought forth wind and rain and lightning, and He shouted and blew until Elephant could take no more and said, "Leave me alone, I will eat trees!"

The Bride Who Became an Elephant

Many years ago…there was an *esiankiki* who was leaving her *enkang* to be married to a man who lived far away. Two men from her future husband's settlement were waiting to take her there. They would have to walk for several days to reach their destination: the two men walking ahead of her, while she followed solemnly and obediently behind. She was tall and beautiful, with young breasts that stood up above her beaded leather skirt. Layers of colorful glass encircled her neck, and strings of beads hung down to her feet.

It was the start of a new life; she had to go, but she did not want to leave her family. Finally she bid them farewell and turned away. She walked past the thorn fence of the enkang and then paused to say goodbye to her mother just one last time. She turned her head, and because of that failing she became an elephant. To this day as a young Maasai girl leaves to be married, she must not turn to make her farewell to her family a second time. This, the Maasai believe, was how the elephants began. It is why the elephants' breasts look like those of a young girl and why the Maasai honor dead elephants in the same way they honor dead humans, by stuffing leaves or long stems of grass into the orifices of their skulls, much like flowers on a grave.

The Pygmies and the Elephant

A MuBira and his wife had a terrible fight. The wife decided to leave and go to her mother's village, but her husband followed her. And as he followed he turned into an elephant. He crashed through the forest and destroyed everything in sight. His wife heard the great noise and climbed a tree. The elephant tracked her and started to knock the tree down. He charged at it and tore at it, but try as he might, he could not pull it down, and his wife held on tightly. Then she called to the pygmies, and the pygmies heard her and came to her rescue. When she saw them she told them to kill the elephant. They attacked it with their spears. They speared it again and again, but it would not die. It charged at the pygmies and drove them back, right back to the BaBira village. There the elephant spoke and said, "You have tried to kill me, your friend, a MuBira." The pygmies replied, "But you have turned into an elephant and tried to kill your wife, and because of that you must cross over to the other side of the river and die." The elephant said, "You want to kill me?" "Yes," said the pygmies, "you have become a very bad elephant." So the elephant went across to the other side of the river. The pygmies went back and took his wife to her mother.

Why Elephants and People Can Never Be Friends

An elephant and a pygmy were friends. The pygmy was called Nbali. He went to visit his friend, the elephant. When he arrived at the elephant's village the elephant was delighted and said to his wife, "My friend Nbali has come, make us a nice dish of mashed plantains." So his wife pounded the plantains, added the salt, and put the dish over the fire. When it was ready the elephant took some red-hot embers and held them to his feet, and the elephant fat ran into the food. When there was enough he turned to his friend, Nbali, and said, "Eat well." The two of them sat down and ate up all the food. The pygmy said, "This is delicious." He went back to his camp. The next day the elephant said to his wife, "Now I will go and visit my pygmy friend, Nbali." So he cleaned himself up and set off. When he arrived at Nbali's camp, Nbali saw him and called, "Welcome, Friend! Come in and sit down." He told his wife that his friend had come, and that she should prepare some mashed plantains for him. His wife prepared the food, and put it on the fire. When it was cooked the pygmy took some red-hot embers and started putting them on his foot. The elephant said, "Don't do that, it will kill you. My feet are big and heavy, let me do it." "Don't be silly," replied the pygmy, and put the embers to his feet. He screamed with pain and almost died. The elephant took hold of him and brought him back to life, then said, "See, I told you it would kill you." The elephant then took the hot embers to his feet so that the oil ran out into the plantains, and they all sat down and ate. When the meal was finished the elephant returned to his village. His wife greeted him and asked if he had had a good time. He said, "It was terrible. My friend Nbali took red-hot embers and put them to his feet and almost killed himself. I shall never go back there again, never."

The Elephant and the Origin of Mankind

Many years ago the elephant came down to earth from heaven, where he was born. He met Lightning, who had also come down. They agreed to hold a competition in noise-making. Elephant started, and he trumpeted so loud that the trees trembled: "Haaah! Hoooh!" But

Lightning just sat there quietly. Then it was his turn. He thundered so loudly that the earth herself shook, the trees were uprooted, their branches broken, and the rivers flooded the country.

The elephant was so frightened that he died on the spot. His body just lay there, and his bowels started fermenting. His stomach began to swell up until it burst, and out of it came all the seeds of all the good plants that Elephant had been eating in Heaven. That is how the vegetables came to earth.

A girl came along one day and found a *tondo* fruit, which is used to cure yaws. It was nicely ripe and red. She took it home and put it in a box. Later, she opened the box again and found a complete man inside. She asked his name, and he said it was Tondo-lindo.

The girl fell in love with him and married him. She gratified all his whims, and they lived in peace for a long time.

One day the man was in the forest and found many seeds and vegetables. It was the place of the dead elephant. He collected them and put them in a box. He came home with it, but no one could open the box. At last an old woman came along and opened it. Lo! Innumerable little children emerged, who flew away in different directions like young ants in the rainy season.

The Origin of Elephants

FRANKLIN EDGERTON

Elephants have never been domesticated. That is to say, they have never been selectively bred to produce a significantly more tractable version of the original wild species. The task may be simply too demanding logistically and, since elephants live approximately as long as humans, time-consuming. But some four thousand years ago, people of the great Indus Valley civilization on the Indian subcontinent learned how to capture and train wild elephants. Hint of that extraordinary accomplishment can be found in the corpus of early Sanskrit literature, including the *Rigveda* (1500–1000 BCE), the *Upanishads* (900–500 BCE), and the great epic *Ramayana*. Another Sanskrit epic, the *Mahabharatha* (with parts composed as early as the ninth and eighth centuries BCE), provides some of the earliest references to elephants used in war. Sanskrit literature also includes minor works that focus particularly on *elephantology*: an organized body of knowledge developed largely in support of elephant management. Even though this is fundamentally a practical lore, Sanskrit elephantology also expresses the cultural and political centrality of elephants through cosmological origin tales such as the one represented in this chapter, which is excerpted from a twentieth-century English translation of *The Matanga-lila (Elephant-Sport) of Nilakantha*.

Since elephants were owned by kings and important for display and war, elephantology was a branch of the science of statecraft, and the

Matanga-lila would have been a treatise supporting the business of the state. According to the translator, Franklin Edgerton, the primary manuscript he worked from was "about two hundred years old," while the work itself, he believed, was likely to be "very much older," conceivably harking back a thousand years or more.

Consisting of 263 stanzas distributed into a dozen chapters, the *Matanga-lila* supports its authority with reference to the mythical founder of elephantology: the sage and "glorious hermit" Palakapya, who was said to have once described his great knowledge to Romapada, King of Anga. It opens, then, with a narrative of the original meeting of King Romapada and Palakapya, which is followed by Palakapya's comments on how the creator god, Brahma, opened the cosmic egg to produce elephants. Originally free from ordinary constraints, those remarkable creatures were able to "assume any shape" and, having wings, they could roam as they wished "in the sky and on the earth." Unfortunately, some of the free-living, shape-shifting creatures fell from grace and came to be "vehicles for even mortal men."

There was an overlord of Anga, like unto the king of the gods, famed under the name of Romapada. Once he was seated on a jeweled throne on the bank of the Ganges in the city of Campa, surrounded by his retinue, when some people reported to him that all the crops of grain, et cetera, were being destroyed by wild elephants. The king reflected: "Now what can I do?"

At this time the distinguished sages Gautama, Narada, Bhrgu, Mrgacarman, Agnivesya, Arimeda, Kapya, Matangacarya, and others, on divine instigation arrived in Campa. The king received them courteously with seats, flowers, and water, et cetera; and out of regard for him they granted the king of Anga a boon, to catch the wild elephants.

On the way the king's men, whom he dispatched to catch the elephants, beheld as they roamed in the jungle a sage, Samagayana, who was staying in a hermitage. Nearby a herd of elephants was grazing; and they saw the glorious hermit Palakapya, who was with the elephant herd, but was separated from it at morning, noon, and night.

All this was reported to the lord of Anga by his servants. So he went and, while the hermit was gone into the hermitage, caught the elephants, came straightaway to Campa, and gave them over to the excellent sages Gautama, Narada, and the rest. But they fastened them securely to posts, and then dwelt there in peace, as did the other folk likewise.

Meanwhile, having performed his service to his father, the hermit Palakapya came out from the hermitage to the place where the elephant herd had been. Not finding it there, he searched everywhere, and so came to Campa, disturbed at heart with affection for them, and tended the elephants in their distress by applying medicines to soothe their wounds, and in other ways.

Now Gautama and the other sages who were there saw this illustrious hermit who was spending his time in silence in the midst of the elephant herd; and so they asked him: "Why do you anoint their wounds? What made you take compassion on the elephant herd?" Though the sages questioned him thus, he made no reply.

Then the noble sages reported these facts, hearing which the king of Anga went thither and paid respects to the hermit with foot-water and other courtesies, and asked him all about his family and name and the rest of his history, being curious to hear. But when that blameless hermit made no reply to him, the king pressed him yet again with questions, bowing low in homage.

Then, propitiated, the sage Palakapya said to the lord of Anga:

"The creation of elephants was holy, and for the profit of sacrifice to the gods, and especially for the welfare of kings. Therefore it is clear that elephants must be zealously tended.

"Of the egg from which the creation of the sun took place, the Unborn Creator took solemnly in his two hands the two gleaming half shells, exhibited to him by the brahmanical sages, and chanted seven *samans* at once. Thereupon the elephant Airavata was born, and seven other noble elephants were severally born, through the chanting.

"Thus eight male elephants were born from the eggshell held in his right hand. And from that in his left hand eight cows were born, their consorts. And in the course of time those elephants, their many sons and grandsons, and so on, endowed with spirit and might, ranged at will over the forests and rivers and mountains of the world.

"And the eight noble elephants of the quarters went to the battle of the gods and demons, as vehicles of the lords of the quarters: Indra, Agni, and the rest. Then in fright they ran away to Brahma. Knowing this, the spirit of *Musth* was then created by Brahma. When it was planted in them, infuriated, they annihilated the host of the demons and went with Indra each to his separate quarter.

"Formerly elephants could go anywhere they pleased, and assume any shape, and they roamed as they liked in the sky and on the earth. In the northern quarter of the Himalaya Mountain there is a banyan tree, which has a length and breadth of two hundred leagues. On it the elephants, after flying through the air, alighted.

"They broke off a branch, which fell upon a hermitage where dwelt a hermit named Dirghatapas. He was angered by this and straightaway cursed the elephants. Hence, you see, the elephants were deprived of the power of moving at will, and came to be vehicles for even mortal men. The elephants of the quarters, however, were not cursed.

"The elephants of the quarters, attended by all the elephant tribes, went and said to the Lotus-born (Brahma): 'O god, when our kinsmen have gone to earth by the power of fate, they may be a prey to diseases, because of unsuitable and undigested food due to eating coarse things and other causes.' Thus addressed by them in their great distress, the Lotus-born replied to them: 'Not long after now there shall appear a certain sage fond of elephants, well versed in medicine, and he shall right skillfully cure their diseases.'

"Thus addressed by Fate (Brahma) the elephants of the quarters went each to his own quarter, while the others, their kinsfolk, went to earth in consequence of the curse.

"A nymph, Rucira, was fashioned by the Creator as he fashioned Speech, by collecting the beauties belonging to sprites, men, demons, and gods. But once she was cursed by Fate (Brahma) because of her evil pride. Hence she was born as a daughter of the tribe of Vasus, from Bhargava, and was named Gunavati. Her great curiosity led her once to the hermitage of Matanga.

"Thinking, 'Nay, she has been sent by Indra to disturb my peace!' Matanga cursed her, and she became an elephant cow. Then the sage, realizing that she was innocent, straightaway said to her, 'Fair elephant cow, when from drinking the seed of the hermit Samagayana a son shall be born to you, then your curse shall come to an end.'

"A certain female sprite once appeared to the hermit Samagayana in a dream. Then the noble hermit straightaway went out from the hermitage and passed water. With the urine, seed came forth. That she drank when the hermit had re-entered the house, and speedily the elephant cow conceived and brought forth a son from her mouth.

"Giving her son with joy to the sage, she left the form of an elephant cow and quickly went to heaven, freed from her curse, in peace. Pleased, the hermit Samagayana then performed the birth rite and other rites for him and in accordance with the instructions of a heavenly voice gave him the name of Palakapya.

"And he played with the elephants, their cows, and the young elephants, roaming with them through rivers and torrents, on mountain tops and in pools of water, and on pleasant spots of ground, living as a hermit on leaves and water, through years numbering twice six thousand, learning all about elephants, what they should and should not eat, their joys and griefs, their gestures and what is good and bad for them, and so forth.

"Know, King of Anga, that I am that hermit Palakapya, son of Samagayana!" Thus addressed by that excellent sage, the King of Anga was greatly amazed.

War Elephants

ARRIAN

War brought elephants to Europe. Indeed, it is entirely possible that the first Europeans of the classical era to see live elephants were Macedonian soldiers in Alexander the Great's army at the start of their Asian campaign: a ferocious sweep of men and horses moving east and intent on conquest. In late September of 331 BCE, Alexander and his army were confronted on the Plain of Gaugamela (in today's Iraqi Kurdistan) by a force commanded by King Darius III of Persia. The Persian army may have included as many as a hundred thousand infantry and cavalry along with some two hundred scythed chariots and fifteen elephants.[5] When, at dawn on the following day, Alexander's army attacked and routed the Persians, however, the fifteen elephants were nowhere to be seen. Darius fled for his life, and the victorious Macedonians finally found the elephants abandoned back in the Persian base camp. Perhaps they had been withdrawn from the battlefield, one commentator suggests, after a fiercely hot day standing in formation and waiting for Alexander to attack.[6]

By early 327 BCE, Alexander was advancing east into the Indian subcontinent, plundering and collecting tribute along the way. In late spring or early summer 326 BCE, he reached the rain-swollen Hydaspes River (now the Jhelum River in the Punjab Province of Pakistan). On the far side of the Hydaspes, King Porus, seven feet tall, resplendent in his armor and seated atop his own elephant, commanded a massive army of infantry, cavalry, chariots, and more than two hundred war elephants, all spread along the riverbank for about two miles and blocking the ford across the swirling Hydaspes. Porus's elephants were armored with fire-hardened leather or quilt-wrapped metal, armed with sharp metal spikes affixed to their tusks, and carrying on their backs wooden castles or platforms supporting archers and javelin throwers.[7]

In the following excerpt, the second-century CE Roman historian Arrian describes the battle of Hydaspes as a spectacular chess game in which Alexander uses subterfuge, psychology, and a superb grasp of tactical dynamics to overwhelm a much larger army. Porus has the advantage not only of superior numbers but also of chariots and those trained war elephants. The chariots were neutralized by muddy ground, however, while the elephants panicked after being directly attacked by the Macedonian infantry until: "Crowded now into a narrow space, the elephants caused as much damage to their own side as to the enemy, turning round and round, barging, and trampling."

Alexander's soldiers had never before faced such a formidable army as that of King Porus, while advance scouts and spies spoke of other Indian kings and their elephant-supported armies ahead, including Xamdrames, who was said to maintain a force that included four thousand war elephants.[8] Dissuaded from continuing deeper into the Indian subcontinent by his own men on the verge of mutiny, Alexander retreated west to establish his headquarters at Babylon, where he ruled from a golden throne and surrounded himself with a personal bodyguard and an elephant corps that, including those taken in battle and those given in tribute, numbered around two hundred. He appointed an *elephantarch*—a commander of the elephants—and he moved to integrate those animals fully into his own army. War elephants would lead Alexander on to further conquests and greater glories...so he must have imagined. Instead he died in the summer of 323 BCE at the age of thirty-two.

Alexander pitched camp on the bank of the Hydaspes, and Porus could be seen on the opposite bank with his entire army and his array of elephants. Porus stationed himself to guard the river at the point where he saw that Alexander had encamped, and posted detachments of guards, each with its appointed commanding officer, to all other stretches of the river where a crossing was easier: his plan was to close off any opportunity for Macedonians to cross. It was the season just after the summer solstice, when there is continuous heavy rainfall over all of India and the melting of snow in the Caucasus, where most of the rivers have their source, greatly increases the volume of their flow. Then in winter the flow reduces, they shrink in size, turn clear, and, except for the Indus and Ganges and it may be one or two others, become fordable in places: the Hydaspes at any rate can certainly be forded.

So Alexander publicly declared that he would wait for this winter season, if he were prevented from crossing for the time being: but he remained as alert as ever on the lookout for a chance to sneak across fast and unobserved. He realized that it was impossible to cross at the point where Porus himself had made his camp, because of the number of elephants there and the large, well-armed, and battle-ready army which would attack his troops as they tried to land. He thought that the horses would refuse even to set foot on the far bank, with the terrifying sight and sound of the elephants coming straight at them, and even before that would not stay on the leather floats for the length of the crossing, but would be spooked by the sight of the elephants ahead and jump into the water. So he made plans to steal across, and this is how he set about it. Night after night he took the bulk of his cavalry up and down the bank, shouting, raising war-cries, and generally making every sort of noise which suggested an army getting ready to cross. Porus would replicate this movement, following the direction of the commotion with elephants and all, and Alexander got him used to this constant tracking. When this had gone on for some time, and nothing happened except shouting and war-cries, Porus stopped responding in parallel to the cavalry excursions, but realized that these were false alarms and stayed put in his camp, though he had set sentries at several points along the bank. Having succeeded in lulling Porus into complacency about these night-time maneuvers, Alexander now put the following stratagem into effect.

At a point where the Hydaspes described a sharp bend, the bank enclosed a projecting loop of land which was thickly wooded with a whole variety of trees: and opposite it there was an island in the river, also wooded and pathless, as it was uninhabited. When he learnt about this island opposite the loop, both places covered with trees and capable of concealing the launch of a crossing, Alexander decided to take his army across at this point. The loop and the island were about seventeen miles from his main camp. Along the whole length of the bank he had posted groups of guards at intervals sufficiently short for them to maintain visual contact with one another and respond readily to orders coming from either direction. At night all along this line they made a great deal of noise and kept fires burning: this continued for many nights.

When Alexander had taken the decision to make the attempt, back at the camp he authorized overt preparations for a crossing. He had left Craterus in charge of the camp with his own cavalry unit, the cavalry from Arachosia and the Parapamisadae, the Macedonian phalanx brigades of Alcetas and Polyperchon, and the western Indian princes with their five thousand men. His instructions to Craterus were that he should not begin to cross the river until Porus had decamped with his army to take on Alexander's force, or else he had learnt that the Macedonians were winning and Porus was in retreat. "But if," Alexander continued, "Porus takes part of his army against me and leaves part behind in the camp with elephants as well, you must still not make a move: if, though, he takes all his elephants with him against me, with some of the rest of his army left in the camp, then cross as fast as you can. It is only the elephants which make it impossible to land the horses—any other part of Porus' army will pose no problem."

Such were his orders to Craterus. Between the island and the main camp where Craterus was left in charge, Meleager, Attalus, and Gorgias had been posted with the mercenary cavalry and infantry: their instructions too were to begin crossing in relays, section by section, as soon as they saw the Indians fully engaged in the battle.

For his own force Alexander selected the elite corps of the Companion cavalry, the cavalry units of Hephaestion, Perdiccas, and Demetrius, the cavalry from Bactria, Sogdiana, and Scythia, and the Dahae mounted archers; from the phalanx the foot guards and the brigades of Cleitus and Coenus; and the archers and Agrianians. He took this force out of sight on a wide detour from the bank, to conceal his approach to the island and the loop, where he had decided to make the crossing. The leather casings for the floats had been stockpiled there for some time, and the work of stuffing them with straw and stitching them tight had been taking place at night. On this particular night a violent rainstorm came on, which helped to smother the preparations for the crossing and the initial moves, with the thunder and the drumming of the rain serving to drown out the clatter of arms and the hubbub of command and response. Most of the boats, including the thirty-oared ships, the triaconters, had been transported to this spot in dismantled sections, put together again out of sight, and hidden in the woods. Towards dawn the wind and the rain had quietened down. Part of the army now began the crossing, the cavalry on the leather rafts and as many infantry as the boats could accommodate: they kept close to the island, so that the scouts posted by Porus would not see them until they had skirted the island and already come near to the bank.

Alexander himself went on board a triaconter for the crossing, together with the bodyguards Ptolemy, Perdiccas, and Lysimachus, and one of the Companions, Seleucus (who later became king), with half of the foot guards in his command. The rest of the foot guards were carried in other triaconters. As soon as this force had passed the island, there was no hiding the final approach to the bank: Porus' scouts saw them coming, and rode off to tell Porus as fast as each man's horse could carry him. Meanwhile a landing was made. Alexander was the first ashore, and took off the foot guards from the other triaconters: with them he arranged the cavalry in immediate formation as more and more disembarked (he had given orders that the cavalry should be the first off the floats), and then took forward this combined force in full battle-order.

What he had not realized was that, in the absence of local knowledge, he had landed not on the mainland but on another island, large enough to be mistaken for the mainland. It was separated from the far bank by a relatively narrow channel of the river, but the violent rain which had lasted for most of the night had swollen the volume of water, so his cavalry could not find the fording-place, and there was the alarming prospect that he would have to repeat the first laborious exercise all over again in order to complete the crossing. In the end the ford was found, and Alexander led the way across, though it was hard going: at the deepest points the water came over the chests of the infantry and the horses had only their heads above the surface. When this final stretch of the river had been crossed, Alexander brought round the elite corps of the cavalry and positioned it on his right wing, together with the best men selected from the other cavalry units, and placed the mounted archers in front of the whole line of cavalry. The infantry were drawn up behind the cavalry: first the king's household guards, commanded by Seleucus, then the royal corps of foot guards, and after them the rest of the foot guards in whatever order the various command units happened to come up at the time. At the wings of the phalanx on either side he placed the archers, the Agrianians, and the javelin-men.

With these dispositions made, Alexander ordered the infantry section, which numbered nearly six thousand, to maintain ranks and follow at marching pace, while he, assuming he had cavalry superiority with the five thousand or so in his force, took the cavalry ahead at speed and unsupported, though he did instruct Tauron, the commander of the archers, to bring his men along behind the cavalry at their own best pace. His plan of engagement was that, if Porus' army met him with their full force, he would either secure a quick and easy victory with a cavalry charge, or at least fight them off until his infantry could join the action. If, though, his unexpected initiative panicked the Indians into flight, he would keep close after them as they ran: the greater carnage he could inflict on the retreating army, the less work would be left for him still to do.

Ptolemy agrees that Alexander first sent his mounted archers against these arrivals, while he himself led out his cavalry in the belief that Porus was approaching with his entire army, and that this cavalry force with Porus' son was the spearhead sent forward in advance of the rest of the army. But when he was given an accurate report of the Indian numbers, he launched an immediate attack on them at the head of his cavalry. They gave way when they saw that Alexander was there in person and that what faced them was not a single line of cavalry but massed squadrons attacking in column. Some four hundred of the Indian cavalry were killed, including Porus' son: the chariots were captured horses and all, as the mud rendered them useless in the actual engagement and weighed them down in the retreat.

When the cavalry who had survived and made their escape reported to Porus that Alexander had succeeded in crossing the river with what amounted to the strongest section of his army, and that his son had died in the fighting, even so Porus was in two minds, as the troops left with Craterus in the main camp opposite could be seen making a start on their own crossing. In the end he chose to move against Alexander and take his whole army into a decisive battle with the strongest part of the Macedonian force and the king himself. Nevertheless he left a few of his elephants behind in the camp with a small contingent of troops, to frighten Craterus' cavalry away from the bank. He himself set out to confront Alexander, taking with him all his cavalry (some four thousand), all his three hundred chariots, two hundred of his elephants, and about thirty thousand serviceable infantry. When he had found a place free of mud, a wide area of sandy soil giving a hard, flat surface for cavalry attacks and turns, he halted there and drew up his army in battle formation.

At the front he placed the elephants in a single long line, at intervals of at least a hundred feet, so that this forward screen of elephants could cover the whole of his infantry phalanx and act as a deterrent to Alexander's cavalry across the entire front. In any case he did not expect that any enemy units would attempt to force through the gaps between the elephants, either on horseback (as horses are frightened of elephants) or, still less, on foot: infantry units would be stopped in a frontal attack by his own hoplites, then trampled as the elephants turned on them. After the elephants he ranged his infantry not quite on the same front, but in a second line slightly behind the beasts, close enough to have the various companies protruding a little way into the gaps. He had infantry also posted on the wings even beyond the line of elephants, and cavalry stationed on each flank of the infantry, with the chariots in front of the cavalry on either side.

This then was Porus' battle-order. When Alexander saw the Indians already taking up formation, he stopped any further advance by his cavalry, to allow the constant stream of infantry to catch up with him. They came to join him at the double, but even when he had the full complement of the phalanx gathered and added to his force, he did not immediately form them up and lead them into battle: they were exhausted and out of breath, and he was not going to commit them in that state to a fresh barbarian army. So he kept his cavalry circling while the infantry could rest long enough to restore their fighting spirit. As he surveyed the Indian battle-line, he decided not to press an attack in the center, where the elephants had been ranged ahead of the front line, with the phalanx drawn up in close order immediately behind and filling the gaps between them: he was cautious of this for the very reasons on which Porus had calculated in making this disposition of his forces. But since he had cavalry superiority Alexander took most of his cavalry with him and rode out to make his attack on the enemy left wing. He sent Coenus, with his own and Demetrius' cavalry units, to the enemy right, with instructions to press close after them in their rear when the barbarians on that wing saw the mass of cavalry attacking them on the left and rode round in support. He had put the infantry phalanx under the command of Seleucus, Antigenes, and Tauron, with orders not to engage until they saw that his cavalry attack had disrupted the enemy infantry line as well as their cavalry.

They were now within missile range, and Alexander sent his mounted archers, about a thousand strong, against the Indians' left wing, to create havoc among the troops stationed there with their dense volleys of arrows and quick charges in and out. And he himself took the Companion cavalry and rode at speed for the enemy left, intent on attacking them in the flank while they were still in disarray and before they could deploy their cavalry in line.

Meanwhile the Indians did indeed concentrate their cavalry from all parts of the field to parallel Alexander's movement and extend their line accordingly: and Coenus' squadrons, as ordered, began to arrive in plain view at their rear. Their appearance forced the Indians to form two fronts, with the larger and strongest part of their cavalry facing Alexander, while the rest wheeled to confront Coenus and his force. This of course caused immediate disruption to the Indian battle-lines and their battle plans. Alexander saw his opportunity, and at precisely the moment when the cavalry were executing this about-turn he attacked those still facing him with such force that the Indians did not even attempt to meet his cavalry charge, but were broken and driven back on the elephants, as if to the protection of a home city wall. At this point the mahouts began to turn their animals against Alexander's cavalry, and the Macedonian phalanx responded with their own attack on the elephants, spearing their riders and, with a ring of men round them, inflicting multiple wounds on the animals themselves. The consequent action was unlike anything they had faced before. The beasts charged into the infantry lines and, wherever they turned, began spreading carnage in the Macedonian phalanx despite the density of its formation. Seeing the infantry in this trouble, the Indian cavalry turned again and made a counter-attack on the Macedonian cavalry: but once more they were penned back towards the elephants, defeated by the far superior strength and experience of Alexander's force. Alexander's cavalry was now amalgamated into a single unit, a formation created naturally by the course of the battle rather than any specific order, and it made repeated charges into the Indian ranks, each time leaving a swathe of slaughter before the Indians could break away. Crowded now into a narrow space, the elephants caused as much damage to their own side as to the enemy, turning round and round, barging, and trampling. The Indian cavalry, tightly corralled among the elephants, suffered massive carnage.

Most of the mahouts had been shot down: wounded, exhausted, and with no one to control them, the elephants could no longer play their specific role in the battle but, maddened by pain, they began attacking friends and enemies alike, crushing, trampling, and killing indiscriminately. The Macedonians, though, had plenty of room and could gauge their attacks on the animals, retreating whenever they charged, but keeping close behind them when they turned and firing javelins at them. Most of the damage done by the elephants was now inflicted on the Indians attempting to rally among them. But as the animals tired and the strength went out of their charges, and all they could do now was trumpet and gradually retire like ships backing water, Alexander threw his cavalry in a cordon round the entire Indian force and gave the signal for his infantry to lock shields and advance the phalanx in the densest possible formation. In the ensuing action all but a few of the Indian cavalry were cut down, and with the Macedonians now pressing the attack on all sides of them there began the slaughter of their infantry also. When a gap opened up in Alexander's cavalry, the survivors all turned and ran.

At the same time Craterus and the other commanders of the army units Alexander had left on the far bank of the Hydaspes began to make their own crossing, once it became clear that Alexander was achieving a decisive victory. Coming fresh to take over the pursuit from Alexander's exhausted troops, they continued to slaughter the retreating Indians on no less massive a scale.

Aristotle's Elephant

ARISTOTLE

Alexander the Great's premature demise was also the demise of a great empire. In a series of campaigns waged across the Middle East, south into Palestine and Egypt, west into Turkey and Greece, Alexander's favored generals and governors fought each other over their various claims and counterclaims, eagerly using the latest and most impressive weapon of war they had ever seen. And in that manifestation, as living battle tanks, elephants were introduced to the classical world of the West. They were comparatively rare to begin with, but as a result of the great arms race following Alexander's death, their numbers were soon bolstered by a restocking of Asian elephants brought overland from India and then, within a generation, African elephants routinely shipped north from Ethiopia on the Red Sea.[9] In Greece, meanwhile, Alexander the Great's death in 323 BCE was followed a year later by the philosopher Aristotle's death at the age of sixty-two.

Aristotle wrote the first detailed descriptions of elephants for Western readers—which raises the question: How did he know so much about them? He was not a world traveler, and he died before the wars initiated by Alexander's successors brought elephants out of Asia and sub-Saharan Africa into the Mediterranean Basin. Aristotle did not, therefore, have any direct experience with a live elephant. Or did he?

One theory holds that he did: that, soon after the defeat of King Darius on the plains of Gaugamela in 331 BC, King Alexander sent to Aristotle in Athens one of Darius's fifteen captured elephants. That particular idea is logistically improbable, but the general concept is not completely absurd. Aristotle's father had served as personal physician to Alexander's grandfather, King Amyntas of Macedonia, and it is not unlikely that Aristotle, as a young boy, had been a playmate to Amyntas's son and Alexander's father, Phillip. After twenty years spent first as a student, and then a teacher, at Plato's Academy in Athens, Aristotle returned to Macedonia's court in 343 and served as tutor to the teenage Alexander.[10] And when the philosopher returned to Athens in 335 BC to establish his own school, the Lyceum, it is reasonable to imagine that King Phillip financed it. After Phillip was assassinated and Alexander became king, it is likely enough that Alexander continued to support Aristotle at the Lyceum.

In any case, much of what Aristotle writes about elephants powerfully suggests that he had information from someone who had observed them, albeit briefly and imperfectly, in natural or seminatural circumstances. Other comments imply that he must have dissected one or had a report from someone who did. Moreover, Aristotle pointedly corrects the errors of other commentators, such as the strange notion expressed by Ctesias (who as court physician to King Artaxerxes of Persia had occasion to speak with traders from India) that the semen of elephants hardens into something resembling amber. We can presume, then, that from somewhere or someone the Athenian philosopher was able to acquire fresh information on elephants.

"Aristotle," biologist and biographer Armand Marie Leroi writes, "was an intellectual omnivore, a glutton for information and ideas. But the subject he loved most was biology. In his works 'the study of nature' springs to life for he turns to describing the plants and animals that, in all their variety, fill our world. To be sure, some philosophers and physicians had dabbled in biology before him, but Aristotle gave much of his life to it. He was the first to do so. He mapped the territory. He invented the science. You could argue that he invented science itself."[11] He also invented natural history. Aristotle's *Historia Animalium* is a supremely ambitious attempt to describe, catalogue, and understand the living world of animals—insects, fish, reptiles, birds, mammals—with a powerful attention to detail based on observation and dissection. Aristotle organized his work not according to species, as an amateur naturalist might, but rather according to groups and systems. As a result, it is not possible to find a chapter on elephants or on any other coherent group at the species level. Instead one finds fragments here and there, vignettes or pieces on elephants that serve to illustrate some larger group or system. Aristotle was a great systematizer, anticipating in that respect the first founder of modern biology, Karl Linnaeus. And he insisted on asking and then answering the *why* questions: Why does this animal behave in such a fashion? Why is that creature built in such a way? His reasoned answers invoke "nature" responding to the environmental circumstances of the organism, which anticipates the thinking of Darwin, modern biology's second founder.

Most of Aristotle's writings on elephants, then, describe them anatomically and are embedded within detailed discussions of comparative anatomy, which is the subject of the first set of excerpts. Aristotle also engages in some less satisfying (sometimes contradictory and often mistaken) attempts to describe natural or seminatural behavior, as seen in the second set. Finally, in the third piece (a single excerpt), we come upon a brief speculation on elephant intelligence and cognition; and of Aristotle's remarks on elephants, this one is both the most provocative and the least sustained by direct observation.

Aristotle's Comments on Elephant Anatomy

The viviparous quadrupeds also have forelegs instead of arms: all the quadrupeds, I say, though the polydactylous ones have something very analogous to hands: at any rate, they used them as hands for many purposes—and we must remember that the limbs on their left side as less independent than in man—with the exception of the elephant: this animal has somewhat indistinctly articulated toes, and its forelegs are much larger than the hind ones. Still, it has five toes, and on its hind legs it has short ankles. Its nose, however, is of such a kind and of such a size that it can be used instead of hands: its method of eating and drinking is to reach with this organ into its mouth, and for its driver...; it even pulls up trees with it, and, when passing through the water, it blows upwards with it. The nose can coil at the tip, but does not bend as a joint, because it contains gristly substance.

Man is the only animal which can actually become ambidextrous.

All animals have a part which is analogous to the chest in man, though dissimilar: in man the chest is broad, in the other animals it is narrow. Only man has breasts in front: the elephant has two, which are near, but not on, the chest.

Apart from the elephant, animals have the flexions of their hind limbs as well as their forelimbs opposite to one another in direction, and to the corresponding flexions in man. Thus, in the viviparous quadrupeds the front limbs bend forwards and the hind ones backwards, and the concavities of the curves are therefore turned to face one another. The elephant does not behave as some used to allege, but settles down and bends its legs, though it cannot on account of its weight settle down on both sides simultaneously, but reclines either on to the left or on to the right, and in that posture goes to sleep. Its hind legs it bends just as a human being does.

Furthermore, in regard to breasts and generative organs, animals differ among themselves and also from man. Some have breasts in front, on the chest or near it, and these have two breasts and two teats as do man and the elephant, as I have said before. The elephant has its two breasts toward the axillae; the female has two quite small breasts, quite out of keeping with the size of its frame, indeed from the side they cannot be seen at all. The males also have very small breasts, like the females.

In some male animals the generative organs are external (examples are man, the horse, and many others), in others internal (example the dolphin). In some of those where the organs are external, they are in front (examples as just quoted); and of these, some have both penis and testicles clear of their body (example, man), others have them closely adhering to the belly, some more, some less so; thus, in the wild boar and the horse this part does not stand clear in the same way. The penis of the elephant resembles that of the horse, though it is disproportionately small for the animal's size; the testicles are not visible externally, being placed within near the kidneys, and for this reason the male quickly frees itself in intercourse. The genital organ of the female [elephant] is situated where the udder is in sheep, and when she is on heat, she draws it back and protrudes it externally, so as to facilitate intercourse for the male; and the organ opens out quite considerably.

The elephant has four teeth on either side, which it uses to masticate its food, grinding it like coarse barley meal, and apart from these it has two others, the "great" ones. In the male these are quite big and turned up at the end; in the female they are small and turned in the opposite direction from those of the male—i.e., they face downwards. The elephant has teeth when it is born, but the "great" ones are not visible at first. It also has a tongue, quite a small one, and placed well inside the mouth, so that it is difficult to see.

The elephant's gut has constrictions, which make it appear to have four stomachs. This gut is where the food has its place: the elephant has no separate receptacle. Its viscera resemble the pig's, except that its liver is four times as large as the ox's, as are its other internal parts, though its spleen is disproportionately small.

On Elephant Behavior

The female elephant begins to submit to the male at ten years of age at the youngest, and never later than fifteen; the male copulates when five or six years old. The season for it is spring. The male will not have intercourse a second time until three years have passed, and never touches again a female it has once impregnated. The period of gestation is two years; the number of young produced at a time, one—another instance of a uniparous animal. The embryo reaches the size of a calf two or three months old.

Male elephants also become savage at pairing-time, and that is why people say that in India elephant-raisers will not allow them to mount the females, the reason being that they get so frantic at this time that they wreck their keepers' houses (jerry-built constructions) and do much other damage. It is also alleged that a lavish supply of food keeps them calmer. In addition, they bring in other elephants to be with them, and chastise and subdue them by setting these others on to belabour them.

The elephant, both sexes, copulates before the age of twenty. After copulation the female carries her young, as some say, for eighteen months; others say for three years. The reason for the disagreement about the period is that it is by no means easy to witness their copulation. The female brings forth by sitting back on her rear parts, and is clearly in considerable pain. As soon as the cub is born it suckles itself by its mouth, not with its trunk. It can walk about and see as soon as it is born.

Elephants too fight fiercely against each other and strike each other with their tusks. The defeated one is strictly enslaved and does not stand against the victor's voice. The elephants in fact differ to a re-markable degree in courage. As war elephants the Indians use the females too, just like the males; however, the females are smaller and much less spirited. The elephant knocks down walls by striking his large tusks against them; and he strikes at palm trees with his forehead until he has felled them, and then by trampling lays them flat on the ground. And the hunting of elephants is done as follows: they pursue them mounted on particular elephants that are tame and brave, and when they have overtaken one they order these tame ones to strike it until they have exhausted it; then the elephant-driver leaps on to its back and directs it with the prong. After this it is quickly tamed and obeys orders. Now while the driver is mounted, they are all gentle; but after he has dismounted, some are and some are not. But they bind the forelegs of those that turn wild with ropes to quiet them down. They hunt them both when they have grown big and as calves.

Some say the elephant lives 200 years, others 120, and that the female lives about the same number of years as the male, and that its prime is about 60 years, and that it does not stand the cold well in the face of winter weather and frosts. The animal lives beside rivers but it is not a river animal. But it makes it way even through water and goes on so long as its trunk reaches above it; for it blows and draws its breath through this. But it cannot swim much, because of the weight of its body.

On Elephant Intelligence and Cognition

The tamest and gentlest of all the wild animals is the elephant, for there are many things that it both learns and understands: they are even taught to kneel before the king. It has quick perception and superior understanding in other respects. After mating with one and making it pregnant it does not touch that one again.

Pliny's Elephants

PLINY THE ELDER

Pliny the Elder, a first-century CE Roman citizen, served as an officer in the Roman army and navy and as an administrator for Roman colonies in Africa and Europe, but he is remembered mainly as the compiler of a thirty-seven-volume *Natural History*. One of the most complete summaries of classical knowledge in existence, this work contains, according to the author's prefatory assessment, twenty thousand "facts worthy of note" winnowed from a library of two thousand books and representing the labors of a hundred expert predecessors. Pliny rather grandly identifies Aristotle as among the most important of them, and the only one with a direct connection to the legendary Alexander the Great:

> King Alexander the Great had a burning desire to acquire a knowledge of zoology, and delegated research in this field to Aristotle, a man of supreme authority in every branch of science. Orders were given to some thousands of people throughout the whole of Asia Minor and Greece—people who made their living by hunting, catching birds, and fishing, as well as those in charge of warrens, herds, apiaries, fish-ponds and aviaries: they were to see that he was informed about any creature born in any region. The result of his inquiries from such people led to the publication, in nearly fifty volumes, of his famous work *On Animals*. I ask my readers to be favourably impressed [by] my presentation of this information—together with facts of which Aristotle was unaware—while making their brief excursion in my direction into all the works of Nature.

On the subject of elephants, Pliny's big advantage over Aristotle is that he lived four hundred years later and thus had a considerable history to draw on, the story of the classical world's contact with live elephants from Asia and Africa.

He could describe, for example, the invasion of Italy by King Pyrrhus of Epirus (northwestern Greek peninsula) in 280 BCE: "the first time," as Pliny has it, that elephants appeared on the Italian peninsula. Italians had never before seen elephants, and undoubtedly their presence in King Pyrrhus's invading army contributed to the rout of Roman legions at the battle of Heraclea. Pyrrhus liberated Tarentum, a Greek city at the southeastern end of Italy, from the Romans; and, with his army strengthened by support from several southern cities in Italy, he advanced to threaten the city of Rome itself. He might have succeeded had not Carthage—that wealthy city of north Africa with colonies in Sicily—formed an alliance with Rome. Pyrrhus turned to attack the Carthaginian colonies in Sicily, and, after several notable successes followed by as many notable failures, retreated across the Adriatic back to Epirus.[12]

Meanwhile, the rulers of Carthage in North Africa had begun developing their own corps of war elephants. Using Egypt as the intermediary, they acquired Indian mahouts and Asian elephants; later they captured African elephants from forests to the south of Carthage. Part of the ordinary training for war elephants was teaching the animals to kill people without hesitation, and the tale recounted by Pliny—concerning the time the famed Carthaginian general Hannibal forced a Roman prisoner to fight an elephant—may have been part of a training program for their elephants.

In any case, the alliance between Rome and Carthage did not last long, and the First Punic War—act 1 in the long and violent drama of Rome versus Carthage—began in 264 BCE. As one important episode in that trans-Mediterranean conflict, Lucilius Metellus, the Roman commander in Sicily, defeated the Carthaginian colonists in Sicily by a clever stratagem that involved trapping and panicking their elephants. Pliny refers briefly to that event, too, and he mentions the additionally clever means by which Metellus managed to ferry some 140 of the captured pachyderms back to Rome, where for a time they served as a spectacle in the circus.[13]

Spectacle was important to the Romans, and—as Pliny tells us— Pompey the Great was involved in at least two of them.[14] Pompey first came into prominence for his role in supporting the Roman general Sulla on the battlefield against General Marius in the Marian-Sullan War. After Sulla marched on Rome and had himself elected dictator in 82 BCE, Pompey secured Sicily against the Marians before sailing to Africa where, in 81 BCE, he defeated rebellious armies there. Sulla, back in Rome, declared his protégé "Great," and the young hero expected a great parade marking his triumphal return to Rome. He arrived at the outskirts of the city with a number of African elephants, and, as a creative addition to the triumph, he yoked together four of them, side by side, and prepared to ride through the city gates in an elephantine chariot. Unfortunately, no one had compared the width of four elephants abreast with the width of the gates, so Pompey was forced to take horses instead.[15]

Pompey's gifts as a soldier and politician brought him to the pinnacle of power in the Roman world. Immensely wealthy and popular as well, he built a great temple to Venus Victrix and in 55 BCE prepared to celebrate its opening with one of the greatest spectacles ever seen. The final day of that event concluded with a battle between elephants and javelin-throwing human gladiators from North Africa. It did not go as planned. As Pliny notes, the wounded elephants tried to break through the iron bars enclosing them, "much to the discomfiture of the spectators." And when the elephants had at last "given up all hope of escape, they played on the sympathy of the crowd, entreating them with indescribable gestures. They moaned, as if wailing, and caused the spectators such distress that, forgetting Pompey and his lavish display specially devised to honor them, they rose in a body, in tears, and heaping dire curses on Pompey, the effects of which he soon suffered."

It is an extraordinary portrait: human cruelty moved to compassion by the actions and cries of tormented animals. Indeed, one of the marks of Pliny's presentation of elephants that distinguishes it from Aristotle's is that Pliny more readily (and uncritically, too often relying on questionable accounts and unsubstantiated hearsay) describes these animals as highly intelligent and emotional. Aristotle was typically more careful to keep his statements within the circle of what was probably known and apparently credible.

The elephant is the largest land animal and is closest to man as regards intelligence, because it understands the language of its native land, is obedient to commands, remembers the duties that it has been taught, and has a desire for affection and honor. Indeed the elephant has qualities rarely apparent even in man, namely honesty, good sense, justice, and also respect for the stars, sun and moon.

Elephants are also credited with an understanding of another's religion, since, when they are on the point of going across the seas, they do not go aboard the ship before being coaxed by their mahout with a sworn assurance about their return. They have been seen, when exhausted by sickness—since diseases assail even those huge bodies—lying on their backs and throwing grass towards the sky as though beseeching Earth to answer their prayers. Indeed, as an example of their docility, they do homage to their king, kneeling before him and offering him garlands. The Indians use a smaller species for ploughing: these they call "mongrel" elephants.

Elephants were harnessed together in Rome for the first time, and drew Pompey the Great's chariot in his African triumph; just as it is recorded that they had been employed on a former occasion, when Bacchus celebrated his triumph after conquering India. Procillus remarks that at Pompey's triumph the yoked elephants were unable to go out through the gates. Some elephants, at the gladiatorial show staged by Germanicus Caesar, even performed clumsy gyrations like dancers.

It was a common trick for them to throw weapons through the air—the wind did not deflect them—and to engage in gladiatorial contests with each other, or to play, together in light-hearted war-dances. Later, elephants even walked tightropes, four at a time, carrying in a litter a woman pretending to be in labor. Or they walked between couches to take their places in dining-rooms crowded with people, picking their way gingerly to avoid lurching into anyone who was drinking.

It is a known fact that one elephant, somewhat slow-witted in understanding orders, was often beaten with a lash and was discovered at night practicing what he had to do. It is amazing that elephants can even climb up ropes in front of them, but more so that they can come down again when the rope is sloping! Mucianus, who held the consulship on three occasions, is the authority for the statement that one elephant learnt the shapes of Greek letters and used to write in Greek: "I myself wrote this and dedicated these spoils taken from the Celts."

Mucianus adds that he had seen elephants at Puteoli, when made to disembark at the end of their voyage, turn round and walk backwards to try to deceive themselves about their estimate of the distance because they were frightened by the length of the gangplank stretching out from the land.

Elephants themselves are aware that their instruments of protection are a valuable commodity that are sought as plunder. Juba calls these "horns," whilst Herodotus, a much earlier source, more appropriately refers to them as teeth, following the commonly accepted term. When these fall out because of some accident or old age, the elephants bury them. Only the tusk is of ivory, otherwise, in these animals too, the skeleton is ordinary bone. Recently, however, because ivory is scarce—outside India an ample supply is rarely found—the elephant's bones have begun to be cut into layers. So much of the ivory in our world has been yielded up to the demands of luxury.

A young elephant is identifiable by the whiteness of its tusks. Elephants exercise the greatest care with their tusks. They use the point of one for fighting, but sparingly to prevent it being blunted, while the other is used for digging out roots and moving large masses. When surrounded by hunters, elephants station those of their number with smaller tusks at the front, so that fighting them is not considered such a challenge. Afterwards, when tired out, they break off their tusks by beating them against trees, and ransom themselves by parting with their booty.

It is strange that most animals know why they are hunted, and almost all of them know what to be on their guard against. An elephant that accidentally encounters a man wandering across its path in some remote place is mild and quiet and even, it is said, points out the way. Yet the very same animal, when it notices a man's footprint, trembles in fear of an ambush before catching sight of the man himself: he stops to pick up the scent, looks about him and trumpets in anger and does not tread on the footprint but digs it up and passes it to the next elephant, and that one to the one following, and so on to the last, with a similar message. Then the column wheels round, retires and forms a line of battle. So much does the scent persist to be smelt by all of the elephants.

Elephants always move in herds. The oldest is the leader, the next in age brings up the rear. When about to cross a river, they send the smallest ahead to prevent the depth being increased by the footsteps of the larger animals wearing away the river bed. Antipater says that two elephants employed by King Antiochus for military uses were known by their names. Indeed elephants know their own names. Although Cato removed the names of the commanders from his *Annals,* he certainly records that the bravest elephant to fight in the Carthaginian army was called "the Syrian," and had one broken tusk.

Antiochus was once trying to ford a river, but his elephant Ajax jibbed, although at other times it was always at the head of the column. So Antiochus announced that the elephant that crossed would have the leading place. Patroclus dared and Antiochus gave him silver trappings (a source of the greatest pleasure for elephants) and all the other privileges of a leader. Ajax, in disgrace, preferred to die of starvation rather than face dishonour. For their sense of shame is remarkable, and when defeated an elephant shuns the voice of its conqueror and offers earth and sacred leaves.

Elephants mate in secret because of their modesty, the male when five years old, the female when ten. This happens for two years in an elephant's life on five days in each year, so men say, and no more. It is no cause for surprise that animals with memory also show affection. Juba records that an elephant recognized, many years later in old age, a man who had been its mahout when young. The same author also cites an example of a sort of insight into justice: King Bocchus tied to stakes thirty elephants he had resolved to punish, and exposed them to the same number of other elephants, while men ran among the latter to provoke them to attack; but they could not be made the instruments of another's cruelty.

Elephants appeared for the first time in Italy during the war with King Pyrrhus, and they got the name "Lucanian oxen" because they were seen in Lucania in AUC 474. They first appeared in Rome, however, five years later, in a triumph. A large number of elephants were captured from the Carthaginians in Sicily by the victory of the pontifex Lucilius Metellus in AUC 502 there were 142, or, as some authorities state, 140, and they were ferried across on rafts which Metellus had made by putting a layer of planks on rows of wine-jars secured together.

Verrius records that these elephants fought in the Circus and were killed by javelins, because the Romans were at a loss what to do with them, since they had decided not to look after them or give them to local kings. Lucius Piso says that the elephants were simply led into the Circus, and, in order to increase the contempt for them, were driven round it by men carrying spears tipped with a ball. The authorities who do not think that they were killed fail to explain their subsequent fate.

There was a famous contest between a Roman and an elephant when Hannibal compelled Roman prisoners to fight one another. He matched a survivor against an elephant and agreed to let him go free if he killed the animal. The prisoner engaged the elephant single-handed and, to the great indignation of the Carthaginians, killed it. Hannibal, realizing that the account of this contest would bring contempt for the beasts, sent horsemen to kill the man as he left the arena.

In Pompey's second consulship, when the temple of Venus Victrix was dedicated, twenty elephants (some say seventeen) fought in the Circus against Gaetulians armed with throwing-spears. One elephant put up a fantastic fight and, although its feet were badly wounded, crawled on its knees against the attacking bands. It snatched away their shields and hurled them into the air. The spectators enjoyed the curving trajectory of their descent—as if they were being juggled by a skilled performer and not thrown by a raging beast. There was also an extraordinary incident with a second elephant when it was killed by a single blow: a javelin struck under its eye and penetrated the vital parts of its head.

All the elephants, en masse, tried to break out through the iron railings that enclosed them, much to the discomfiture of the spectators. (For this reason, when Caesar as dictator was intending to stage a similar spectacle, he surrounded the arena with a moat filled with water. The Emperor Nero did away with this moat when he added seats for the knights.) But when Pompey's elephants had given up hope of escape, they played on the sympathy of the crowd, entreating them with indescribable gestures. They moaned, as if wailing, and caused the spectators such distress that, forgetting Pompey and his lavish display specially devised to honour them, they rose in a body, in tears, and heaped dire curses on Pompey, the effects of which he soon suffered.

Elephants also fought on the side of Julius Caesar the dictator in his third consulship: twenty were matched against five hundred infantry and, on a second occasion, a further twenty, equipped with howdahs, each carrying sixty men, fought with the same number of infantry as previously and an equal number of cavalry. Later, during the principates of Claudius and Nero, elephants fought men in single combat; for gladiators this was the high point of their career.

Captured elephants are very quickly tamed by barley juice. In India they are rounded up by a mahout, who, riding a tame elephant, either catches a wild one on its own, or separates one from the herd and beats it so that when it is exhausted he can mount it and control it in the same way as the tame one. Africans employ covered pits to trap elephants. When one strays into a pit the rest of the herd immediately heap branches together, roll down rocks and build ramps, using every effort to drag it out. Previously, in order to tame elephants, the kings herded them with the aid of horsemen into a trench constructed by hand and deceptive in its length. Enclosed within its banks the beasts were overcome by starvation. The proof of submission was when an elephant gently took a branch that a man held out to it. Nowadays hunters pierce their feet, which are very soft, with javelins in order to obtain their tusks.

Elephants, once tamed, are used in war and carry howdahs full of armed soldiers on their backs. In the East they make a major contribution to warfare, scattering battle-lines and trampling down armed men. Yet these beasts are terrified by the slightest shrill sound made by a pig. When wounded and frightened, they always yield ground and cause no less destruction to their own side. African elephants fear their Indian counterparts and do not dare to look at them, for the Indian species is larger.

People commonly think that elephants carry their unborn offspring for ten years, but Aristotle says two years: they produce only one at a time, and they live for two hundred, in some cases three hundred years. They reach maturity at sixty. They particularly enjoy rivers and roam around streams, although unable to swim because of the size of their bodies and because they cannot bear cold; this is their greatest weakness.

Elephants eat with their mouth, but breathe, drink and smell with their trunk, not inappropriately called their "hand." They hate mice most of all living creatures, and if they see one even touch the food put in their stall they back away from it. They experience extreme pain if, when drinking, they swallow a leech (which I observe has now begun to be commonly called a "bloodsucker"). This fixes itself in the windpipe and inflicts unbearable pain.

Their tusks command a high price and the ivory is excellent for images of the gods. Our extravagant life-style has found another reason for singing the praises of the elephant, namely the taste of the hard skin of the trunk, sought after for no other reason than that one seems to be eating ivory itself. Large tusks are seen in temples. Polybius, on the authority of Prince Gulusa, also records that in remote parts of Africa having a common border with Ethiopia tusks are substitutes for doorposts in houses, and that in these and in cattle stalls, partitions are made with elephants' tusks in place of stakes.

India produces the biggest elephants, as well as snakes that continually fight them. The snakes are of such a size that they easily surround the elephants in coils, and tie them up with a twisted knot. In this struggle both die, for the defeated elephant falls and its weight crushes the snake coiled round it.

Beasts of the Book

T. H. WHITE

Aristotle wrote of the "character" of animals and that they seem to possess "a certain natural capability in relation of each of the soul's affections—to intelligence and stupidity, courage and cowardice, to mildness and ferocity, and the other dispositions of this sort."[16] But generally he avoided speculating about the realm of the invisible. He reported what he saw, experienced, or reasonably believed based on rational assessments of the work of others. He was a professional skeptic, in other words, among the first in the tradition of scientific skepticism. Pliny lacked that discipline, and as a result he seldom hesitated to fill the black box of the unknown with any interesting fancy that came to mind. "Indeed," as he writes in the opening paragraph of his chapter on elephants, "the elephant has qualities rarely apparent even in man, namely honesty, good sense, justice, and also respect for the stars, sun, and moon."[17] Really?

Knowledge is a kind of signal: hardly different from a single tiny light flashing Morse code or a grand broadcast of radio waves rippling from earth to outer space. Like any other signal, knowledge will degrade over time unless it is regularly refreshed and refocused. A particular body of knowledge—elephantology, for instance—can be refreshed through new observations and refocused through a discipline of skeptical assessment and experimental testing. I'm speaking, then, of scientific method and a scientific culture: things that would not begin to emerge in recognizable form until the end of the European Renaissance.

Western elephantology in the classical era was an inheritance from ancient authors who knew, or thought they knew, something about elephants. It was a body of knowledge that came from long-standing Eastern traditions and from the centuries of contact that occurred after living elephants were brought into the Mediterranean basin for use in war and entertainment. The fading of that European knowledge happened partly because its founder and first champion, Aristotle, was sui generis, a lone scientist working in a prescientific world, and partly because—after the fall of Rome and the end of most productive European exchanges with Africa and Asia—there were no elephants to see. And because there were no elephants to see, no European would produce fresh and original observations of them for hundreds of or even perhaps a thousand years, so elephantology became entirely a bookish discipline. One turned to the library shelf. One read and repeated what the great authors had themselves read and repeated from earlier great authors. And the quality of this body of knowledge, its accuracy and legitimacy, became remarkably degraded over time.

The Book of Beasts is an astonishing book, a medieval bestiary and a wonderful sample of medieval art, scholarship, and thinking. It is also a good example of what happens when knowledge is transmitted through time without the support of fresh observation, steady skepticism, or astute experimentation. Scholars trace the book's origin to an anonymous author nicknamed "the Physiologus," who may have written in Greek and probably lived in Egypt at some time between the second and fifth centuries CE. In writing his book, the Physiologus referred to the usual authorities of his time and place: Herodotus, Ctesias, Aristotle, Pliny, and so on. He was, like Pliny, a dedicated compiler of the best knowledge available, and he concentrated on animals: the beasts of his bestiary.

Unlike Pliny's *Natural History*, though, the Physiologus bestiary was not an encyclopedia about animals for a single audience from a single era and civilization. It was more of a Wikipedia, a compilation of entries that traveled through time and space and in the process grew and changed through an accretion of contributions and emendations from many, many unknown contributors. "Perhaps no book except the Bible," declares the American linguist E. P. Evans, "has ever been so widely diffused among so many people and for so many centuries as the Physiologus. It has been translated into Latin, Ethiopic, Arabic, Armenian, Syriac, Anglo-Saxon, Spanish, Italian, Provençal, and all the principal dialects of the Germanic and Romanic languages."[18] This *Book of Beasts* was regarded as a precious piece of true knowledge about the world's domestic, wild, strange, and fantastical creatures. Scholars and scribes and translators spread the bestiary, and as they did so they added new meanings through the inevitable warp of transcription and translation, and new materials—new beasts—whenever they chose. The early Greek version presented forty-nine different beasts, while the Latin version that circulated in northern England during the twelfth century CE contained between two and three times that number. A copy of that Latin version, apparently preserved at the Lincolnshire abbey of Revesby before it came into the care of Cambridge University, is the source of the English translation presented here.

Like Pliny, the Physiologus recognized the great authority of Aristotle, yet he presents elephants and the other animals as if they are psychological and cognitive beings. That is a reasonable assumption, but without the discipline of scientific skepticism and research it quickly leads to an unreasonable anthropomorphizing.

Aristotle had written, at one point, that elephant females have a gestation period lasting, "some say," eighteen months—while "others say for three years."[19] "The reason for the disagreement about the period is that it is by no means easy to witness their copulation." Aristotle's statement that uncertainty about the length of gestation results from elephant behavior—in other words, that elephants behave in such a way that matings are seriously difficult to observe—seems to be a source for Pliny's unwarranted presumption that "elephants mate in secret because of their modesty."[20] And a thousand years later, in the Latin *Book of Beasts*, that same unwarranted presumption comes to be presented as an amazing and critical zoological fact. Sexual modesty is now a central psychological feature of elephants, one of the major reasons why God created them in the first place: to serve as living types or exemplars of the Christian story and moral virtues.

─── ── ── ── ── ── ── ───

There is an animal called an elephant which has no desire to copulate.

People say that it is called an Elephant by the Greeks on account of its size, for it approaches the form of a mountain: you see, a mountain is called *eliphto* in Greek. In the Indies, however, it is known by the name of *barrus* because of its voice—whence both the voice is called *bari*tone and the tusks are called ivory (*ebur*). Its nose is called a pro-boscis (for the bushes), because it carries its leaf-food to its mouth with it, and it looks like a snake.

Elephants protect themselves with ivory tusks. No larger animals can be found. The Persians and the Indians, collected into wooden towers on them, sometimes fight each other with javelins as if from a castle. They possess vast intelligence and memory. They march about in herds. And they copulate back-to-back.

Elephants remain pregnant for two years, nor do they have babies more than once, nor do they have several at a time, but only one. They live three hundred years. If one of them wants to have a baby, he goes eastward toward Paradise, and there is a tree there called Mandragora, and he goes with his wife. She first takes of the tree and then gives some to her spouse. When they munch it up, it se-duces them, and she immediately conceives in her womb. When the proper time for being delivered arrives, she walks out into a lake, and the water comes up to the mother's udders. Meanwhile the father-elephant guards her while she is in labor, because there is a certain dragon which is inimical to elephants. Moreover, if a serpent happens by, the father kills and tramples on it till dead. He is also formidable to bulls—but he is frightened of mice, for all that.

The Elephant's nature is that if he tumbles down he cannot get up again. Hence it comes that he leans against a tree when he wants to go to sleep, for he has no joints in his knees. This is the reason why a hunter partly saws through a tree, so that the elephant, when he leans against it, may fall down at the same time as the tree. As he falls, he calls out loudly; and immediately a large elephant appears, but it is not able to lift him up. At this they both cry out, and twelve more elephants arrive upon the scene: but even they cannot lift up the one who has fallen down. Then they all shout for help, and at once there comes a very Insignificant Elephant, and he puts his mouth with the proboscis under the big one, and lifts him up. This little elephant has, moreover, the property that nothing evil can come near his hairs and bones when they have been reduced to ashes, not even a Dragon.

Now the Elephant and his wife represent Adam and Eve. For when they were pleasing to God, before their provocation in the flesh, they knew nothing about copulation nor had they knowledge of sin. When, however, the wife ate of the Tree of Knowledge, which is what the Mandragora means, and gave one of the fruits to her man, she was immediately made a wanderer and they had to clear out of Paradise on account of it. For, all the time that they were in Paradise, Adam did not know her. But then, the Scriptures say: "Adam went in to his wife and she conceived and bore Cain, upon the waters of tribulation." Of which waters the Psalmist cries: "Save me, O God, for the waters have entered in even unto my soul." And immediately the dragon subverted them and made them strangers to God's refuge. That is what comes of not pleasing God.

When the Big Elephant arrives, i.e. the Hebrew Law, and fails to lift up the fallen, it is the same as when the Pharisee failed with the fellow who had fallen among thieves. Nor could the Twelve Elephants, i.e. the Band of the Prophets, lift him up, just as the Levite did not lift the man we mentioned. But it means that Our Lord Jesus Christ, although he was the greatest, was made the most Insignificant of All the Elephants. He humiliated himself, and was made obedient even unto death, in order that he might raise men up.

The little elephant also symbolizes the Samaritan who put the man on his mare. For he himself, wounded, took over our infirmities and carried them from us. Moreover, this heavenly Samaritan is interpreted as the Defender about whom David writes: "The Lord defended the lowly ones." Also, with reference to the little elephant's ashes: "Where the Lord is present, no devil can come nigh."

It is a fact that Elephants smash whatever they wind their noses round, like the fall of some prodigious ruin, and whatever they squash with their feet they blot out.

They never quarrel about their wives, for adultery is unknown to them. There is a mild gentleness about them, for, if they happen to come across a forwandered man in the deserts, they offer to lead him back into familiar paths. If they are gathered together into crowded herds, they make way for themselves with tender and placid trunks, lest any of their tusks should happen to kill some animal on the road. If by chance they do become involved in battles, they take no little care of the casualties, for they collect the wounded and exhausted into the middle of the herd.

COLONIAL
AND INDUSTRIAL
ELEPHANTS

Killers and Heroes

ROUALEYN GORDON-CUMMING

Within a century after the Portuguese explorer Vasco da Gama rounded the Cape of Good Hope in 1497, the European colonial powers had begun their program of exploration and exploitation in Africa. Particularly in southern Africa, European sport hunters discovered a paradise of large and exotic wild animals to kill, and the result was catastrophic. From an estimated one hundred thousand elephants roaming free in southern Africa at the start of the seventeenth century, fewer than a hundred individuals remained by 1920.[21]

The worst of the European hunters were not furtive, anonymous individuals looking for a little profit from selling ivory. Rather, they were bold and capable men who wrote books, gave lectures, donated stuffed trophies to museums, and were feted for their exploits in the European capitals. They became heroes, in short—men who presented themselves to the world as tough adventurer-explorer types and called themselves sportsmen, although what they seemed to love above all was simply the sport of spilling blood: the exciting act of killing big animals. They reveled in the gore of their gory trade.

Roualeyn George Gordon-Cumming was one of the killer-heroes from the Victorian era. The second son of a Scottish baronet and educated at Eton, Gordon-Cumming spent two years in India as an officer in the Madras Light Cavalry before arriving in southern Africa in 1843 to ride for a year with the Cape Mounted Rifles. By 1844, however, he had set out, accompanied by a group of African assistants—trackers and porters and gun bearers—into what was then the uncharted interior, regions of southern Africa known as Bechuanaland and the Limpopo River Valley. Gordon-Cumming was a big man with a big red beard, and he hunted grandly on horseback while dressing boldly in a traditional plaid kilt. The area he moved through was, at the time, rich in wildlife, and he killed as he moved. He was interested in ivory, of course—who was not?—but he also looked for trophies. After five years of the hunting life in Africa, he returned to Britain to assemble his stuffed trophy collection and write his memoir, *Five Years of a Hunter's Life in the Far Interior of South Africa*, which was published in 1850. The trophy collection, which he called the South African Museum, was exhibited in London at the Great Exhibition of 1851.

In his memoir, Gordon-Cummings demonstrates his mastery of deft understatement and the subtle boast—as when, for example, he describes in an offhanded couple of sentences how he killed a charging white rhino and then, after rising late the following day, had a satisfying breakfast of "coffee and rhinoceros." He also seems to have mastered the aristocratic arts of impervious equanimity and cool indifference to the needs or suffering of others, including "the tallest and largest bull elephant I had ever seen" (a "noble" animal), whose killing he prolongs in order to enjoy a cup of coffee, review a few pleasant memories and philosophical contemplations, and then partake in a bit of experimental target practice at the slowly dying giant.

On the 31st, I held southeast in quest of elephants, with a large party of the natives. Our course lay through an open part of the forest, where I beheld a troop of springboks and two ostriches, the first I had seen for a long time. We held for Towannie, a strong fountain in the gravelly bed of a periodical river: here two herds of cow elephants had drunk on the preceding evening, but I declined to follow them;

and presently, at a muddy fountain a little in advance, we took up the spoor of an enormous bull, which had wallowed in the mud, and then plastered the sides of several of the adjacent veteran-looking trees. We followed the spoor through level forest in an easterly direction, when the leading party overran the spoor, and casts were made for its recovery. Presently I detected an excited native beckoning violently a little to my left, and, cantering up to him, he said that he had seen the elephant. He led me through the forest a few hundred yards, when, clearing a wait-a-bit, I came full in view of the tallest and largest bull elephant I had ever seen. He stood broadside to me, at upward of one hundred yards, and his attention at the moment was occupied with the dogs, which, unaware of his proximity, were rushing past him, while the old fellow seemed to gaze at their unwonted appearance with surprise.

Halting my horse, I fired at his shoulder, and secured him with a single shot. The ball caught him high upon the shoulder-blade, rendering him instantly dead lame; and before the echo of the bullet could reach my ear, I plainly saw that the elephant was mine. The dogs now came up and barked around him, but, finding himself incapacitated, the old fellow seemed determined to take it easy, and, limping slowly to a neighboring tree, he remained stationary, eyeing his pursuers with a resigned and philosophic air.

I resolved to devote a short time to the contemplation of this noble elephant before I should lay him low; accordingly, having off-saddled the horses beneath a shady tree which was to be my quarters for the night and ensuing day, I quickly kindled a fire and put on the kettle, and in a very few minutes my coffee was prepared. There I sat in my forest home, coolly sipping my coffee, with one of the finest elephants in Africa awaiting my pleasure beside a neighboring tree.

It was, indeed, a striking scene; and as I gazed upon the stupendous veteran of the forest, I thought of the red deer which I loved to follow in my native land, and felt that, though the Fates had driven me to follow a more daring and arduous avocation in a distant land, it was a good exchange which I had made, for I was now a chief over boundless forests, which yielded unspeakably more noble and exciting sport.

Having admired the elephant for a considerable time, I resolved to make experiments for vulnerable points, and, approaching very near, I fired several bullets at different parts of his enormous skull. These did not seem to affect him in the slightest; he only acknowledged the shots by a "salaam-like" movement of his trunk, with the point of which he gently touched the wound with a striking and peculiar action. Surprised and shocked to find that I was only tormenting and prolonging the sufferings of the noble beast, which bore his trials with such dignified composure, I resolved to finish the proceeding with all possible dispatch; accordingly, I opened fire upon him from the left side, aiming behind the shoulder; but even there it was long before my bullets seemed to take effect. I first fired six shots with the two-grooved, which must have eventually proved mortal, but as yet he evinced no visible distress; after which I fired three shots at the same part with the Dutch six-pounder. Large tears now trickled from his eyes, which he slowly shut and opened; his colossal frame quivered convulsively, and, falling on his side, he expired. The tusks of this elephant were beautifully arched, and were the heaviest I had yet met with, averaging ninety pounds weight apiece.

Industrial Killers

W. D. M. BELL

The invention in 1798 of a mechanical ivory saw by the Connecticut clockmaker Phineas Pratt brought the Industrial Revolution of the West to the elephants of Africa.

Pratt's mechanical saw was hand-powered and represented a minor improvement over the traditional, back-and-forth hand-sawing methods a neighbor already used in his own cottage business of converting elephants' teeth to combs for hairdressing. But the new saw was also circular, and thus it readily attached to a rotating mechanical power source—the torque provided by a windmill or water mill. Within a short while, Pratt's invention was producing some 250 ivory combs a day. Within a few decades, the Pratt family business and a few other ivory cutting factories located in the lower Connecticut River Valley were mass producing several ivory items for the American market, including combs, letter openers, business cards, dominoes, cuff links, collar buttons, billiard balls, and piano keys.[22]

Ivory for piano keys found the biggest market. As precision machinery began replacing skilled labor in their manufacture, pianos became increasingly affordable for aspiring members of the American middle class. In a single decade, from 1850 to 1860, American manufacturers expanded their production of pianos from 9,000 to 22,000 per year; by the start of World War I, American factories were producing well over a third of a million pianos annually.[23] All those pianos required keyboards and keys, and the ideal keys presumed a surface, as the essential interface between a player's sensitive fingers and the key's insensate block of wood, made from a polished slice of elephant tooth. Each piano keyboard required about a pound and a half of elephant tooth, and the descendants and associates of Phineas Pratt owned the factories that produced the majority of those sliced teeth. Between 1851 and 1869, the company known as Pratt, Read & Co. maintained a near monopoly on the American market, selling every year around a half million dollars' worth of ivory for piano keys and items.[24]

By then, most ivory coming to America was shipped from a single East African port located on the coastal island of Zanzibar. Arab traders during the previous thousand years had established the connections and infrastructures for the highly profitable ivory trade, which was commonly conjoined with the similarly profitable slave trade. An American who visited Zanzibar wrote back home in 1844 that the merchants of Zanzibar could acquire ivory for almost nothing—a few beads or a length of brass wire in exchange for a large elephant's tooth. Moreover, the writer continued, "It is the custom to buy a tooth of ivory and a slave with it to carry it to the sea shore. Then the ivory and slaves are carried to Zanzibar and sold."[25]

Among the most notorious of the slave traders was a man named Hamed bin Muhammed—informally known as Tippoo Tib—who operated out of Central Africa. In 1882, the missionary Alfred Swann met Tippoo Tib's caravan headed east for Zanzibar and described it thus: "As they filed past we noticed many chained together at the neck. Others had their necks fastened into the forks of poles about six feet long, the ends of which were supported by the men who preceded them. The neck is often broken if the slave falls in walking. The women, who were as numerous as the men, carried babies in addition to a tusk of ivory on their heads."[26]

European colonial powers ended the legal slave trade in East Africa by the end of the nineteenth century, but the commercial trade in mass-produced ivory products, which financed the deaths of between a quarter of a million to one million elephants per decade, continued well into the twentieth. It was ordinary for European "sportsmen" of the period to shoot all the elephants they could, sometimes providing their African assistants with extra guns in order to add bodies to the bag, then consuming by the campfire a few favored pieces, such as the heart, a piece of trunk, a foot, before hacking out the ivory. Meanwhile, of course, the killing technology Europeans brought to the continent continued to improve, while the professional hunters—an Englishman named W. D. M. Bell comes to mind here—actively developed their killing skills.

Bell was scientific about it. He dissected an elephant's skull in order to study the location of the brain in relation to prominent marks on the head's exterior, such as the eyes and ear holes, so that he could master "the head shot." He crawled inside an elephant's carcass, instructing African assistants to thrust spears through the body at various points in order to understand the best location and angle to reach the vital parts—heart, lungs, arteries—for what he called "the body shot." But the head shot was certainly preferable, Bell wrote, and among the many advantages was that "it causes instantaneous death, and no movement of the stricken animal communicates panic to others in the vicinity. The mere falling of the body from the upright to a kneeling or lying position does not appear in practice to have any effect other than to make the others mildly curious as to what has happened." By comparison, a bullet to the heart means that the stricken elephant "almost invariably rushes off with a groan and squirm for fifty or a hundred yards, taking with him his companions, which do not stop when he stops, but continue their flight for miles."

Perhaps Bell's most productive hunting experience took place in the Lado Enclave. The region was once part of the Egyptian province of Equatoria and a center for the ivory and slave trade along the west bank of the Upper Nile. It was taken by the British in 1869 and administered by Sir Samuel Baker, who suppressed the slave trade. But the British, planning a railroad that would cross the continent from Cairo to the Cape, then negotiated with the Belgian king Leopold II to exchange a significant bit of his property, in eastern Congo, for a less significant bit of theirs in what, once the treaty was signed in 1894, was named the Lado Enclave. The treaty leased Lado to King Leopold for the rest of his life plus six months, after which it would revert to British control. In 1908, Bell arrived in the Lado Enclave and acquired a "limitless" hunting permit covering five months for £20. With the support of sixty African assistants, Bell slaughtered 210 elephants in Lado and carried away five tons of ivory.[27] In the following excerpt, taken from his 1923 memoir *The Wanderings of an Elephant Hunter*, Bell describes part of his first year at Lado—which seems to be "a perfect game paradise."

———————————————————

At the time of which I write, the Enclave de Lado comprised the country bordering the western bank of the Upper Nile from Lado on the north to Mahazi in the south. It was leased to Leopold, King of the Belgians, for the duration of his life and for six months after his death. This extension of the lease was popularly supposed to be for the purpose of enabling the occupiers to withdraw and remove their gear.

While the King was still alive and the Enclave occupied by the Congo authorities, I stepped ashore one day at Lado, the chief administrative post in the northern part. Luckily for me, the Chef de Zone was there, and I immediately announced my business, the hunting of elephant. The Chef was himself a great shikari, and told me he held the record for the (then) Congo Free State, with a bag of forty-seven, I think it was. He was most kind and keenly interested in my project, and promised to help in every way he could.

As regards permission to hunt, he told me that if I merely wished to shoot one or two elephants, he could easily arrange that on the spot, but that if I wanted to hunt elephant extensively, I should require a permit from the Governor, who lived at Boma, at the mouth of the Congo. The price of this permit was £20, and it was good for five months in one year. It was quite unlimited and, of course, was a gift to anyone who knew the game. The Belgians, however, seemed to think that the demanding of 500 frs. for a permit to hunt such dangerous animals was in the nature of pure extortion; they regarded as mad anyone who paid such a sum for such a doubtful privilege.

I was, naturally, very eager to secure such a permit, especially when the Chef told me of the uncountable herds of elephants he had seen in the interior. By calculation it was found that the permit, if granted, would arrive at Lado in good time for the opening of the season, three months hence. I deposited twenty golden sovereigns with the Treasury, copied out a flowery supplication to the Governor for a permit, which my friend the Chef drafted for me, and there was nothing more to do but wait.

Having entered my rifles at Lado and cleared them through the Douane, it was not necessary again to visit a Belgian post. So when the hunting season opened, I already had a herd of bull elephant located. Naturally, I lost no time when the date arrived. The date, that is, according to my calculations. This matter is of some importance, as I believe I was afterwards accused of being too soon. I may have begun a day or even two days before the date, but to the best of my knowledge it was the opening date when I found a nice little herd of bulls, several of which I killed with the brain shot. I was using at that time a very light and sweet-working Mann.-Sch. carbine, .256 bore and weighing only 5¼ lb. With this tiny and beautiful little weapon I had extraordinary luck, and I should have continued to use it in preference to my other rifles had not its Austrian ammunition developed the serious fault of splitting at the neck. After that discovery I reverted to my well-tried and always trusty 7 mm. Mauser.

My luck was right in on that safari. The time of year was just right. All the elephant for one hundred miles inland were crowded into the swamps lining the Nile banks. Hunting was difficult only on account of the high grass. To surmount this one required either a dead elephant or a tripod to stand on. From such an eminence others could generally be shot. And the best of it was the huge herds were making so much noise themselves that only a few of them could hear the report of the small-bore. None of the elephant could be driven out of the swamps. Whenever they came to the edge and saw the burnt-up country before them, they wheeled about and re-entered the swamp with such determination that nothing I could do would shake it. Later on when the rains came and the green stuff sprang up everywhere—in a night, as it were—scarcely an elephant could be found in the swamps.

Just when things were at their best, tragedy darkened the prospect. Three of my boys launched and loaded with ivory a leaky dug-out. The leaks they stopped in the usual manner with clay, and shoved off. The Nile at this place was about a mile broad, and when about half way the entire end of the canoe fell out; it had been stuck in with clay. Down went everything. Now, here is a curious thing. Out of the three occupants two could swim. These two struck out for the shore and were drawn under by "crocs," while the one who could not swim clung to the barely floating canoe, and was presently saved. He kept trying to climb into it from the side, which resulted, of course, in rolling it over and over. It may well be that to this unusual form of canoe maneuvering he owed his life. At any rate, the crocs never touched him. A gloom settled on the camp that night certainly; but in Africa, death in, to us, strange forms produces but little impression on the native mind, and all was jolly again in a day or two.

After about two months' hunting it became necessary to bury ivory. The safari could no longer carry it, so a site was chosen close to the river bank and a huge pit dug. Large as it was, it barely took all our beautiful elephant teeth. Ivory has awkward shapes and different curves, and cannot be stowed closely. Consequently there was much earth left over after filling the pit, to show all and sundry where excavating had taken place. Where cattle or donkeys are available, the spot is enclosed by a fence of bushes, and the animals soon obliterate all traces, but here we had nothing. So to guard our precious hoard I erected a symbol which might have been mistaken by a white man for a cross made in a hurry, while its objectless appearance conveyed to the African mind the sure impress of "medicine." I remember that to one rickety arm I suspended an empty cartridge and the tip of a hippo's tail. That the medicine was good was shown three or four months later when I sent some boys for the ivory. They found that the soil had been washed away from the top, exposing completely one tusk and parts of the others, but that otherwise the cache was untouched. In spite of my giving to the boy in charge of this party one length of stick for each tusk contained in the pit, he returned with one short of the proper number. To convince him of this fact it was necessary to line out the ivory and then to cover each tusk with one bit of stick, when, of course, there remained one stick over. Straightway that party had a good feed and set off for the pit again, well over a hundred miles away. It never occurred to them that one tusk might have been stolen. They were right. Through just feeling that they themselves would not have touched anything, guarded as our pit was guarded, they judged correctly that no other native man would do so. At the bottom of the open pit and now exposed by the rains was found the missing tusk.

After the hot work of the dry season in the swamps the open bush country, with still short grass, was ideal for the foot-hunter. The country was literally swarming with game of all sorts. I remember in one day seeing six white rhino besides elephant, buffalo and buck of various kinds. Then happened a thing that will sound incredible to most ears. I ran to a standstill, or rather to a walking pace, a herd of elephant. It happened thus. Early one morning I met with a white rhino, carrying a magnificent horn. I killed him for the horn. At the shot I heard the alarm rumble of elephant. Soon I was up to a large herd of bulls, cows, half-growns, and calves. They were not yet properly alarmed, and were travelling slowly along. Giving hasty instructions to my boy to find the safari and to then camp it at the water nearest to the white rhino, I tailed on to that elephant herd. The sun then indicated about 8 a.m., and at sundown (6 p.m.) there we were passing the carcass of the dead rhino at a footpace. By pure luck we had described a huge circle, and it was only by finding the dead rhino that I knew where I was. Throughout that broiling day I had run and run, sweating out the moisture I took in at occasional puddles in the bush, sucking it through closed teeth to keep out the wriggling things. At that time I was not familiar with the oblique shot at the brain from behind, and I worked hard for each shot by racing up to a position more or less at right angles to the beast to be shot. Consequently I gave myself a great deal of unnecessary trouble. That I earned each shot will become apparent when I state that although I had the herd well in hand by about 2 p.m., the total bag for the day was but fifteen bulls. To keep behind them was easy, the difficulty was that extra burst of speed necessary to overtake and range alongside them. The curious thing was that they appeared to be genuinely distressed by the sun and the pace. In the latter part of the day, whenever I fired I produced no quickening of the herd's speed whatever. No heads turned, no flourishing of trunks, and no attempted rushes by cows as in the morning. Just a dull plodding of thoroughly beaten animals. This day's hunting has always puzzled me. I have attempted the same thing often since, but have never been able to live with them for more than a short distance. Although a large herd, it was not so large but that every individual of it was thoroughly alarmed by each shot. I think that perhaps they committed a fatal mistake in not killing me with a burst of speed at the start. I left them when I recognized the dead rhino, and found camp soon afterwards. The next two days I rested in camp, while the cutting-out gang worked back along the trail of the herd, finding and de-tusking the widely separated bodies of the dead elephants.

Shortly after leaving this camp four bull elephants were seen in the distance. As I went for them, and in passing through some thick bush, we came suddenly on two white rhino. They came confusedly barging about at very close range, and then headed straight for the safari. Now, it is usual for all porters familiar with the black rhino to throw down their loads crash bang whenever a rhino appears to be heading in their direction. Much damage then ensues to ivory if the ground be hard, and to crockery and bottles in any case. To prevent this happening I quickly killed the rhino, hoping that the shots would not alarm the elephants. We soon saw that they were still feeding slowly along, but before reaching them we came upon a lion lying down. I did not wish to disturb the elephants, but I did want his skin, which had a nice dark mane. While I hesitated he jumped up and stood broadside on. I fired a careful shot and got him. He humped his back and subsided with a little cough, while the bullet whined away in the distance. At the shot a lioness jumped up and could have been shot, but I let her go. Then on to our main objective.

WORKING
AND PERFORMING
ELEPHANTS

To Break and Tame

U TOKE GALE

In the nineteenth century, Britain maintained its colonial empire through the reach of a great navy. Teak wood, exceptionally resistant to insects and rotting, was critical to marine construction, and the best teak in the world grew wild in Burma. Unfortunately, prime wild teak in Burma was found only in the remote and mountainous forests of the northwest, which were almost completely inaccessible for machinery. Thus, the British drafted elephants.

Extracting teak in Burma was done in twenty-five- to thirty-year rotational cycles. Prospectors would locate the largest tree in an area, kill it by cutting a ring in the bark at the base, then allow the still-standing tree to dry out for the next three years. Drying out was essential given that green teak will sink in water, and the timber had to be transported to sawmills and export markets by floating it on the rivers. After the drying period, loggers came to fell the selected trees using hand tools. They then cut the felled timber into thirty-foot sections, cutting in each a fist-sized hole at one end and threading through that hole a heavy chain. Elephants, hitched to the chains with thick woven-fiber harnesses, would drag the great logs—potentially equivalent to their own body weight—down to a dry stream or river bed. Dragging timber was their main task, but the animals would also push massive logs with their trunks and heads, or grab and lift it, and they would periodically toss those enormous pieces of hardwood off a high cliff and down to a dry stream or river bed hundreds of feet below, where it would stay cached for the remainder of the dry season. The elephants also hauled people and supplies through the mountains.

When the monsoons came, the dry stream and river beds turned wet. As the water rose, it lifted and floated the teak downstream and out to larger rivers and then the main waterways, the Chindwin and Irrawaddy Rivers. There the logs were lashed together into huge rafts and guided by steamboat to sawmills and timber markets at Mandalay and Rangoon. Of course, it was hard to predict how high the streams and rivers would rise in any given year and which logs would be successfully floated downriver, but elephants helped launch and guide the logs. Once they were afloat, the logs frequently became snagged and jammed in raging currents, but elephants would be called upon to enter the rushing water and, playing an intricate and dangerous game of giant pick-up sticks, unjam the logs.[28]

For the most part, the elephants used in this work were wild, living freely in the forests of Burma and elsewhere in Southeast Asia. They were transformed into captive and working ones using traditional techniques developed thousands of years earlier.

One method of capture, described in U Toke Gale's classic memoir *Burmese Timber Elephant* (1974), involved driving a herd of wild elephants—with shouting, horns, drums, fire crackers, fires, and smoke—into a very large stockade anchored to standing trees and consisting of horizontal poles woven with cane and rope to vertical posts twelve to sixteen feet high, a foot or more in diameter, and sunk three to four feet into the ground. Some of the trapped herd, particularly large and aggressive males, might be shot. Others were released. Vulnerable and malleable young elephants were most often kept for breaking in, taming, and training.

Breaking in amounted to a sudden and severe lesson in helplessness. Using already trained and fully mature elephants, men would drag the screaming, thrashing, and kicking young animals out of the stockade and immobilize them with ropes into a cradle: one strong tree, two stout posts, and an overhead cross beam. For the next two or three days, the immobilized and desperate animal would be weakened by semistarvation until, "its spirit more or less broken," two skilled trainers would start the taming process, which involved gradually—over the another ten to twenty days—offering small amounts of water and food and occasional treats (such as bamboo leaves, banana leaves, rice, and salt-and-tamarin balls) to the captive, while more or less continuously chanting to the animal from dawn until far into the night and gently touching him or her all over. Periodically, one of the trainers would climb onto the overhead beam and then, hanging from the beam, stand on the animal's back for a few minutes. The breaking-in process was undoubtedly harrowing, devastating enough that in "extremely rare cases" elephants would kill themselves by stepping on their trunk ends and thus cutting off their own air supply. The animal would be "so bent on suicide that no amount of shouting, swearing and spearing with sharpened sticks of bamboo by the trainers could, in any way, scare him into removing his foot or relaxing its pressure on the trunk." The taming part must not have been much better.

Once sufficiently tamed, the elephant would be at last released from the crush and allowed to forage around camp, kept from wandering too far by fetters attached to his or her forelegs, and subjected to three months of initial training. After that, mahouts (known in Burma as *oozies*) and their apprentices (*pejeiks*) took on the important job of managing animals who soon enough would weigh a few tons and stand seven and a half to nine feet high at the shoulders. Oozies would perch confidently at the neck of an elephant and then persuade—with a verbal command, a barefoot kick into sensitive areas behind the ears, a poke or whack with an ankus or bullhook—the otherwise indifferent giant to move forward, to turn right or left, to kneel with the hind legs and then with the front, to grasp and lift mightily, to push and pull and drag massive boles of timber from one place to another.

One might wonder why the elephants put up with it, but, of course, they had no choice. Meanwhile, in exchange for a hard day's work they were given afternoons and evenings off (though often fettered and sometimes dragging a chain), predictable access to food, protection from predators, attentive veterinary care, the occasional treat of tamarin-and-salt balls, a stable social world, and a few weeks of rest and relaxation each year during the hottest season. As creatures of memory, passion, and intent, elephants are more dangerous than machines, but they can do important things that machines so far cannot. They can move confidently into difficult and precarious places, and they can haul and lift enormous objects with surprising finesse. They can also make quick and creative decisions when unexpected things happen.

Elephants still work at logging in the mountains of Burma (renamed Myanmar in 1989), although the use of machinery has increased, so that much of the floating of timber down to the sawmills has been replaced by transporting overland by truck. And while the selective logging and the use of elephants have contributed to a vision of "sustainable" harvesting in this part of the world, other factors, such as widespread corruption and a vast illegal trade, have reduced the nation's forest cover since independence in 1948 by a third. As a result, the government recently instituted a ban on the export of raw logs and simultaneously slashed its total logging quotas. This move, intended to protect the remaining forests and hopefully to produce significant gains in employment for people in the wood-processing industry, is also expected to cut total demand for logged timber by as much as four-fifths. That reduction in demand will reduce the need for the approximately 5,500 elephants in Myanmar still valued as skilled labor. Where will they go? If released into the forest, they might survive as feral animals in a limited environment. Or perhaps they will be slaughtered for their ivory and hides or smuggled into Thailand and inducted into the abusive business of amusing tourists.[29]

U Toke Gale's *Burmese Timber Elephant* describes an astonishing world of people and elephants that may seem, in our contemporary moment, antique and that is probably vanishing. In the following excerpt from that book, the author describes the initial techniques—of breaking and taming—used to transform a wild elephant into a working one.

Soon after the capture, the elephant trainers start roping in the captives, an undertaking which definitely calls for dexterity, cool courage, patience and a lot of stamina. Men sometimes go without food from dawn to dark trying to bind one foreleg or a hindleg, or the neck of a captive, as the case may be, with strands of manila or *shaw* ropes, one end of each secured to large trees or strong pegs driven deep into the ground for this purpose.

The captive, screaming and kicking wildly, and lashing out its trunk to the left and to the right, is then gradually dragged and removed, as reported earlier, with the aid of a full-grown tamed elephant. It takes from ten to twenty days, or at most a month, to tame a youngster below the age of twenty years. I have seen young calves who could be hand-fed four or five days after the capture. I have before now inspected full-grown elephants that were captured just four months ago, and submitted to my inspection prior to purchasing them as though they were born in captivity. On the other hand, some full-grown bulls, having submitted to training, may develop into dangerous man-killers, and names like Tun Tin, Ngwe Maung (Silver Gong), Doh Lone, Shwelayaung (Golden Moonlight), and Ko-Laung-Kyaw, among others, jump to my mind as typical cases of captured bulls, who at the slightest provocation would kill their attendants, or anyone who absent-mindedly ventured within reach of their trunks. Before the war in 1942, the last named tusker—a *swai-gar*—had killed nine people, as his name implies, but by the time he came under my charge in the Kawlin-Wuntho area during the Japanese occupation, he had proudly added another three victims to his glorious record of killings. The three latest victims were women—"bazaar-sellers" as we called them—all killed within a few minutes of each other, and one had her lower limbs dangling from a branch of a tree some twenty feet above the ground, and right over a dirt-road used by villagers between two large villages.

Extremely rare cases of strangulation have been unofficially reported among adult captives of both sexes in the early stages of breaking-in process. The strangulation is self-inflicted, and in most instances, it is brought about by the animal pressing his forefoot down on his trunk. He is so bent on suicide that no amount of shouting, swearing and spearing with sharpened sticks of bamboo by the trainers could, in any way, scare him into removing his foot or relaxing its pressure on the trunk.

The animal under training is denied food or water for two or three days. At the end of this period, its spirit is more or less broken, and from this stage onwards, it comes under the care and tuition of two men whose main duty is to fetch and carry whenever necessary adequate fresh fodder and clean water for their trainee. Usually, some rice and salt wrapped in a fresh, green plantain leaf is offered twice a day, and which it consumes with relish. The two men are also expected to keep the cradle clean of all droppings and stale fodder each morning and evening. At nights, they sleep in turns. One of the cardinal points to remember at this stage of education, is to see that the animal is kept awake—morning, noon and night—with very little food to eat, and hardly any drink at all. Sleep would restore strength,

and strength in a newly captured wild elephant is by no means con-
ducive to training.

The two trainers constantly talk to their "pupil," saying sweet
nothings, and frequently touching it lightly on the rump, thighs,
stomach and shoulders. One man would now and then stand, for a
few minutes at a time, on its back, balancing himself by holding on
to the beam placed above the cradle. And all the time, the two men
keep talking from dawn into the night—softly, loudly, endlessly in
a sing-song voice, addressing it repeatedly as "my little brother, my
big brother," or, "my little sister, my big sister," depending on the sex
and age of the pupil under training. The secret of effective training
lies in the wise application of the principle of the carrot-and-stick
technique, more of the carrot—or, as in the case of the Burmese elephant—more, perhaps, of ripe tamarind and salt—than the big stick.

After ten or twenty days of such humane treatment, the animal is
docile enough to be released near the camp with cane fetters fastened
on its forelegs. Cane fetters are normally used when training captives,
as they are easier to handle than the iron ones, and cheaper by far.
Some trainers would fasten the front leg to one of the hindlegs with a
shaw rope, kept fairly slack so as to allow a certain amount of freedom
of movement, but at the same time, to see that the animal cannot take
a stride much longer than two feet. A small wooden *khalauk* is hung
from its neck. It is always a good policy to reemploy as *oozie* and *pejeiks* the same men who have had their hand in training the animals,
so that the understanding which had imperceptibly developed at the
cradles or crushes may be projected into their new life together as a
team carrying out the intricate business of extracting timber.

In another three months' time, [the captive elephant] has learnt to
remember the sounds of about a dozen words of command, such as
myauk (lift), *hmet* (lie down), *pway* (carry), *ya* and *how* (be careful, be
quiet). At the appropriate word of command, it will pick up from the

ground small sticks of firewood; lift one of its forelegs to enable the *oozie* to climb on to its head; and go down on its knees and elbows at the time of daily baths in a jungle stream. Some Hkamti Shans pierce a hole in the lobe of one of the ears, and a soft, manila cord of about a quarter of an inch thick is then inserted through it for use as a leash in the latter stages of its training, when it would be led by the cord for baths or short walks along the sandy banks of a stream.

Before long, the captive proudly takes its place among a team of six or eight trained baggage elephants, with an *oozie* on his neck, and with whom it will probably work for another fifty years or so. When it is about eighteen years of age, it will have learnt to push teak logs with its forehead, break up log-jams in a jungle stream, and drag small logs precariously along narrow and slippery drag-paths.

But as in all forms of education, its training never really ends, and it is only through the slow and strenuous process of experience that it will eventually understand all the intricacies of timber extraction, and the danger to which it is always exposed. It is very seldom that a well-trained elephant can be classed as a very good working animal until it has attained the age of twenty-five or thirty years.

A Mother's Love

J. H. WILLIAMS

Born in Cornwall, England, in 1897, James Howard Williams served in the British Camel Corps during World War I. After the war, he searched for a civilian occupation that might combine his twin passions for animals and adventure, which brought him to apply for work as a forestry assistant in Burma. He started in 1920, moving to a four-hundred-square-mile section of forest in the remote mountains of northwest Burma where ten logging camps had been established to cut and move teak with the help of seventy elephants. His job was to move nomadically from camp to camp, checking on the workers and their elephants.

At the start of World War II, with the threat of a Japanese invasion imminent, all the wives and children of British timber employees in Burma were ordered to evacuate, but by the time the evacuation was underway the Japanese were already bombing Rangoon, Burma's only significant port. Soon the official evacuation included large numbers of Indian and Burmese refugees, and a desperate trek across the rugged northwestern mountains into eastern India was led by Williams and more than a hundred elephants and their oozies.

Then the British were driven out of Burma by the Japanese army, and Williams, driven by his love for elephants, while also recognizing that the Japanese would seize and exploit them for their own strategic purposes, crossed back into Burma on foot. Working as a Special Operations officer from hidden camps in the mountains, he created the British Army's first and only Elephant Company. He made contact with many of the oozies who had taken their elephants into hiding during the Japanese advance, and eventually, with a full force of working elephants at hand, using the animals to build roads and construct hundreds of bridges in circumstances that were otherwise impossible, he assisted in the retaking of Burma. He settled back in England at the start of 1946 and began his second career as a memoirist based on his earlier adventures.

James Williams predated U Toke Gale by almost a generation, having begun his career in the Burmese teak industry eighteen years earlier. But the two men shared a similar education on the nature of elephants, the kind that comes from intimate, dirty, sweaty, dangerous, daily contact. Both men describe elephants as surprisingly and creatively intelligent creatures. And both seem to agree that among their most remarkable qualities is an expressive emotionality, one emphatically demonstrated in the powerful bonds between mothers and their young—"a mother's love," as Williams describes it in this brief excerpt from his memoir, *Elephant Bill*.

Elephants are good swimmers and extremely buoyant. When the oozie is going to cross a large river, such as the Chindwin or the Irrawaddy, with his elephant, keeping it under control, he fits a surcingle under its belly and over the withers, kneels on the animal's back and grips the rope in front of him, using a small stick, instead of his feet, to signal his "aids," behind its ears. In this position he is on top of the highest point of the elephant.

Once they are under way and in deep water, it is most amusing to watch. For a time the elephant will swim along gaily, with a rather lunging action. Then, all of a sudden, the oozie will snatch a deep breath, as his mount goes down, like a submarine, into fifteen feet of water. The animal, for pure fun, will keep submerged, almost to bursting point, trying to make his rider, who goes down with him, let go.

But the oozie knows that an elephant can only stay under water for the same length of time as a man. So he holds on. The elephant, meanwhile, is doing a fairy-like dance on tiptoe along the bottom, while the poor old oozie is wondering if he will ever surface. Suddenly both reappear, blowing tremendously and taking great gasps of breath.

In crossing a wide river, where the elephant has to swim a thousand yards or so, he may drift as much as four hundred yards downstream. He does not make any strenuous effort to make the crossing where the river is narrowest, or to reach a particular point on the opposite bank.

One evening, when the Upper Taungdwin River was in heavy spate, I was listening and hoping to hear the boom and roar of timber coming from upstream. Directly below my camp the banks of the river were steep and rocky and twelve to fifteen feet high. About fifty yards away on the other side, the bank was made up of ledges of shale strata. Although it was already nearly dusk, by watching these ledges being successively submerged, I was trying to judge how fast the water was rising.

I was suddenly alarmed by hearing an elephant roaring as though frightened, and, looking down, I saw three or four men rushing up and down on the opposite bank in a state of great excitement. I realised at once that something was wrong, and ran down to the edge of the near bank and there saw Ma Shwe (Miss Gold) with her three-months-old calf, trapped in the fast-rising torrent. She herself was still in her depth, as the water was about six feet deep. But there was a life-and-death struggle going on. Her calf was screaming with terror and was afloat like a cork. Ma Shwe was as near to the far bank as she could get, holding her whole body against the raging and increasing torrent and keeping the calf pressed against her massive body. Every now and then the swirling water would sweep the calf away; then, with terrific strength, she would encircle it with her trunk and pull it upstream to rest against her body again.

There was a sudden rise in the water, as if a two-foot bore had come down, and the calf was washed clean over the mother's hindquarters and was gone. She turned to chase it, like an otter after a fish, but she had travelled about fifty yards downstream and, plunging and sometimes afloat, had crossed to my side of the river, before she had caught up with it and got it back. For what seemed minutes, she pinned the calf with her head and trunk against the rocky bank. Then, with a really gigantic effort, she picked it up in her trunk and reared up until she was half standing on her hind legs, so as to be able to place it on a narrow shelf of rock, five feet above the flood level.

Having accomplished this, she fell back into the raging torrent, and she herself went away like a cork. She well knew that she would now have a fight to save her own life, as less than three hundred yards below where she had stowed her calf in safety, there was a gorge. If she were carried down, it would be certain death. I knew, as well as she did, that there was one spot between her and the gorge where she could get up the bank, but it was on the other side from where she had put her calf. By that time, my chief interest was in the calf. It stood, tucked up, shivering and terrified on a ledge just wide enough to hold its feet. Its little, fat, protruding belly was tightly pressed against the bank.

While I was peering over at it from about eight feet above, wondering what I could do next, I heard the grandest sounds of a mother's love I can remember. Ma Shwe had crossed the river and got up the bank, and was making her way back as fast as she could, calling the whole time—a defiant roar, but to her calf it was music. The two little ears, like little maps of India, were cocked forward, listening to the only sound that mattered, the call of her mother.

Any wild schemes which had raced through my head of recovering the calf by ropes disappeared as fast as I had formed them, when I saw Ma Shwe emerge from the jungle and appear on the opposite bank. When she saw her calf, she stopped roaring and began rumbling, a never-to-be-forgotten sound, not unlike that made by a very high-powered car when accelerating. It is the sound of pleasure, like a cat's purring, and delighted she must have been to see her calf still in the same spot, where she had put her half an hour before.

As darkness fell, the muffled boom of floating logs hitting against each other came from upstream. A torrential rain was falling, and the river still separated the mother and her calf. I decided that I could do nothing but wait and see what happened. Twice before turning in for the night I went down to the bank and picked out the calf with my torch, but this seemed to disturb it, so I went away.

It was just as well I did, because at dawn Ma Shwe and her calf were together—both on the far bank. The spate had subsided to a mere foot of dirty-colored water. No one in the camp had seen Ma Shwe recover her calf, but she must have lifted it down from the ledge in the same way as she had put it there.

Jumbomania: A Circus Story

P. T. BARNUM

The Sanskrit *Matanga-lila* identifies three ways of "driving" elephants: with words, feet, and the hook. Words can be used in three ways: caressingly, neutrally, and abusively. Using the feet or the hook, though, requires physical contact with an elephant at his or her most vulnerable points, particularly areas behind the ears and on the head. "Inserting both his feet in the neck chain, with strokes of his firmly implanted toes and heels, and with his thighs tightly binding," we read, a mahout "shall take a good hold of his hook in his right hand, and with the other hand likewise a staff of reed or the like, and with concentrated mind shall gently make the elephant go by indications made with his feet" emphatically applied to sensitive spots at the backs of the elephant's ears.

Brahma created the elephant hook—or ankus—in four different shapes: that of a thunderbolt, a half-moon, a spike, and a thorn. All four kinds of elephant hook will be applied to six vulnerable places on an elephant's head in six distinctive ways: "barely touching, pressure, hard striking, pressure after brandishing the hook, pulling back after brandishing, and...violent swinging around after brandishing." Thus states the *Matanga-lila*, identifying the ancient and traditional ways of controlling an elephant.[30]

Ropes, chains, fetters, and hooks are still used to control working elephants in the mountains of Myanmar. We know from photographs, film, and written accounts, moreover, that the same or similar restraints and devices have been used to control circus elephants in the West during the nineteenth and twentieth centuries, and into the twenty-first. Indeed, from an elephant's perspective, working in an Asian logging camp must be only mildly distinguishable from working in a Western circus. The confinement, restraint, stress, pain, frustration, and social alienation must be about the same for both; the main difference is that circus elephants perform in front of a human audience that has paid to be amused. Amusement of that sort is a cotton-candy emotion: an evanescent sense that the world is a light and sweet and frothy place where trouble is absent and pleasure abounds. To maintain such a mood for the audience requires concealing the dark parts—for example, the instruments of control that too closely resemble instruments of torture—and creating a surface of light parts, a drama of bright lights, bright music, clowns, popcorn, and animals doing silly things in silly costumes. That's the positive image and the implied story that every circus has sought to maintain.

Jumbo, perhaps the most famous circus elephant of all, was captured in Africa and sold to the Paris Zoo, which in 1865 sent him to the London Zoo in exchange for a rhinoceros. Jumbo was about four years old and five feet high at the shoulders when he arrived in London and was presented as a children's pet: sweet, docile, and ready to play his part as an adventure ride. He was soon big enough to carry on his back a balanced pair of wooden benches, arranged back to back and stable enough to carry a half dozen or more riders facing in opposite directions. Meanwhile, a woman sold treats—rolls, cakes, fruit—for zoo-goers to feed their pet. According to a journalist writing for the *Spectator*, Jumbo realized the "popular ideal" for elephants, given that he resembled "a moving mountain." Children especially were "delighted with him, for after they get down [from a ride], they feel they have done something brave and risky."[31]

By 1880 Jumbo, approaching adulthood, became subject to periodic rages, perhaps in response to the arrival of musth. He smashed his
head and feet against the walls and doors of his barn, which were then
reinforced with heavy timbers and thick steel plates. He pierced the
steel plates with his tusks, and in the process both tusks were shattered
to the root. By December 1881, the zoo's superintendent, recognizing
this animal as a genuine danger to the public, purchased an elephant
gun. Jumbo was then around eleven feet tall at the shoulders, and he
weighed some six thousand pounds, which is in the normal range for
a male African savanna elephant his age. If he followed the normal
pattern, he would continue to grow for the next two to three decades,
and thus the danger of a catastrophic event, a rampaging monster instead of a children's pet, was likely to get worse. Fortunately, a brazen
American huckster and showman named Phineas T. Barnum, looking
for new human and animal exhibits to entice even larger crowds to his
museum and traveling circus, offered to buy Jumbo for £2,000.

After the offer was accepted, the problem became a physical
one: getting him out of the London Zoo and onto a ship bound for
America. On February 17, 1882, Barnum's agents arrived at the zoo
with ten horses hitched to an oak-and-iron crate on a steel-wheeled
trolley and proceeded to wrap the elephant in chains. At the last minute, he foiled his new captors by lying down inertly in front of the
crate—and at this point an additional problem emerged. The British
public, roused mightily by the vision of a noble giant in chains and then
goaded by the patriotic press, began a series of passionate responses
that altogether constituted a national mania, dubbed Jumbomania.
The sale of Jumbo was debated in the House of Commons and was
protested by Queen Victoria and the Prince of Wales. A legal challenge
was launched by a rebel faction of Fellows of the London Zoo. The
editor of Vanity Fair opened a Jumbo Defence Fund. Cartoonists drew
cartoons. Composers composed songs. Children wrote sweet letters,

while grownups penned threatening ones. Others sent him candy, biscuits, fruit, a large pumpkin, twelve dozen oysters, a wedding cake, beer, wine, whiskey, flowers, locks of hair, a gold necklace, a sewing machine, a large box of seasickness pills, a very large nightcap to keep his head warm. But although the London Zoo's managing council was often the target of this public passion, in the end the zoo profited substantially. Ticket sales rose fivefold, then tenfold, until on Jumbo's last day at the zoo, March 20, 1882, some 18,500 paying visitors showed up, hoping to wish fond farewell to the great British beast.

That was the story, and on the other side of the Atlantic Jumbo's new owner, Phineas T. Barnum, gloried in all the free publicity. He encouraged it, participated in it. Ultimately, Jumbomania, as it spread to America, meant Barnum was able to recoup within three weeks of increased circus income all the expenses involved in buying the animal and shipping him to New York. Of course, the story had little to do with Jumbo, who was still only a traumatized animal imprisoned in a cage, yet whatever dishonesty was required or inspired by the taking of Jumbo from the London Zoo was eagerly expanded by Barnum in America, who always maintained an amateur's interest in truth-telling.

The following excerpt, written as a publicity press release by Barnum or one of his anonymous publicists a few months after Jumbo arrived in America, describes in predictably exaggerated terms the tale of Jumbo's removal from zoo life in England to begin his circus life in America.

No period of the past has ever furnished a sensation like Jumbo. If anybody has anything stupendous, no matter what nature the enterprise, undertaking or object, it is at once styled Jumbo. Jumbo means a big thing now-a-days. It is a new and popular word for everybody's everyday lexicon, and it is not specially confined to Jumbo himself. Jumbo also means success of a Herculean order, and delight, instruction and recreation to the millions everywhere, besides certain destruction to all pretenders and would-be competitors, who foolishly get under his feet or within sound of his trumpet-trunk. Six million Americans saw and admired him in the Eastern States last year, and almost as many more attempted, but failed to get an eye on him. The rush was greater than could be accommodated.

Jumbo is said by the people and the press to be a Feature that more than fills the bill. His mastadonic size and really mastadonic shape overwhelm the beholder. And he has grown several inches, his tusks have pushed out nearly a foot, and he weighs a full ton more than last year. He is gradually increasing his already monstrous and phenomenal development. There never was but one Barnum—there is but one Jumbo. They are inseparable, and irrevocably the property of history, and *the* products of the century—and each so familiar to the reader of the current events of the age that further description here is unnecessary. Say "Jumbo"—and there is a volume of comment in silence, because it is the one accepted synonym in all the world for magnitude.

Eighteen years ago, the Royal Zoological Garden, Regent's Park, London, England, purchased from the Jardin des Plants, Paris, the Elephant "Jumbo," of the African species, and an account, of apparent authenticity, says: "When he arrived from Paris he was a wee little pachyderm, not more than five feet high; but he has since attained a height over twelve feet, and weighs nearly ten tons. He is already more huge in his proportions than the monster that Ptolemaeus states was brought by Caesar to Britain in 54 B.C., and terrified the inhabitants greatly; and he is more colossal than the 'elephant of enormous size' which was presented by the King of France to our own Henry III, in 1252."

Sir Samuel Baker says he was acquainted with the history of Jumbo's earlier existence. He asserts that he was captured when very young by the Hamran Arabs, who brought him down from the Settite River in Abyssinia, and disposed of him to a Bavarian collector named Johann Schmidt. Jumbo was then less than four feet high, and traveled with another elephant about his own tender age, which has since died. Even if this is true, and there is no reason to dispute it, there is still no accurate data upon which to make a calculation as to his age. He was obtained at the Jardin des Plantes, Paris, in 1861, and after years of rapid growth was transferred to the Royal "Zoo," London. As zoologists have learned from close observation that the African species of elephant grows almost imperceptibly for many years of its youth, requiring a much longer time to attain maturity than its Indian kindred, it will be seen how impossible to guess rationally at Jumbo's age. He is probably much older than generally believed.

Jumbo's prodigious size has been the wonder of all Europe for years. It is probable that half a million Americans saw and admired him during his sojourn at the Royal London "Zoo." It is estimated by the managers of the "Zoo" that in seventeen years a million and a quarter of children have ridden gleefully on his broad back, including the most noted among the royalty of England and other European powers. He was a gigantic favorite with everybody, and was fed by his friends with barrels of buns, precious bits and sweet morsels every day.

P. T. Barnum, and his associates, J. A. Bailey and J. L. Hutchinson, have for two years had argus eyes on the elephantine monster, and a trusted agent was sent to England to purchase him, if possible. The Council, in whose hands the business of the Zoological Society is placed, was called together by the Superintendent, Mr. Bartlett, and Barnum, Bailey and Hutchinson's agent told them if they would set a price on Jumbo, that he would convey their determination to his employers. After consultation and much delay, they named the enormous sum of TWO THOUSAND POUNDS AS THEIR PRICE FOR THE GIANT BEAST. When the information was conveyed to the managers by the agent, they concluded to accept the terms, and at once communicated their resolve to the Council of the Zoological Society of England. Preparations were inaugurated for the removal of Jumbo to America. It was found that the hatchways of the largest ship would have to be cut away, and the upper deck raised in order to admit the great brute, which many well-informed persons believe to be the mastodon of bygone ages. This difficulty surmounted, the agents, with a corps of experienced elephant keepers, at once set sail for England. In the meantime, the press had taken up the theme on the other side of the Atlantic, and there went up such a remonstrance and resistance against the sale and removal of the favorite Jumbo, as the world has seldom heard of before. Hundreds of cables were received by the great showman and his partners, offering almost any consideration if they would cancel the sale. Tearful requests from thousands of children and anxious ladies appeared in the columns of the British public prints. Her Majesty, the Queen, and the Prince of Wales joined the entreaties and regrets with the anguish of the whole of England. The illustrated newspapers published serious pictures and ludicrous cartoons of poor Jumbo, and the craze swept with the rapidity of lightning. Hundreds of columns and editorials appeared in the leading press daily, and the weeping nation poured in wagon loads of protesting letters by every mail. Letter heads, cards, ladies' charms, umbrellas, fans, hats, bonnets, boots, gloves, overcoats, etc., were got up in London with the portrait of Jumbo on them. His picture was sold by the hundreds of thousands on the street. The Royal Zoological Gardens added nearly $50,000 to its receipts from those who wanted to make a final farewell visit to Jumbo. The excitement spread even to the House of Commons, was discussed in Parliament and every section of Great Britain. The people, through the London *Daily Telegraph*, offered Mr. Barnum any sum in reason to cancel the contract and permit Jumbo to remain. The children wrote him thousands of letters, offering to pay him any amount if he would leave them in possession of their dear old favorite, Jumbo.

Every obstruction possible to invent was put in the way of the agents and keepers by the people, who censured no party to the transaction save the directors of the Zoological Society. Queen Victoria and the Prince of Wales requested the Garden to refuse to deliver Jumbo, and let Barnum collect his damages, which the British nation would pay. Mr. Ruskin did the same thing. Our Ambassador, Mr. Lowell, said in public speech: "The only burning question between the nations is Jumbo." Mr. Laird, builder of the rebel war steamer *Alabama,* wrote Mr. Barnum a letter in like tone and manner. To all these appeals Mr. Barnum gave courteous attention but turned a deaf ear. He was firm as Gibraltar, and stood boldly and courageously for his rights, answering the London *Daily Telegraph* by cable: "Fifty millions of American citizens awaiting Jumbo's arrival. My forty years' invariable practice of exhibiting the best that money could procure, makes Jumbo's presence here imperative. Hundred thousand pounds would be no inducement to cancel purchase."

Among the elephants in the Garden is a female called Alice, which, from the affection which has many years existed between the two, has been known as Jumbo's wife. When the Americans appeared and chained the huge monarch, preparatory to taking him through the streets to Millwall Docks, nine miles away, he utterly refused to leave his companion, making the saddest demonstrations of grief. At the gate he laid down, and no power of the keepers could compel him to move further. After days of ineffectual attempts, it was at last decided to build a huge trolley, or truck, with low wheels, the weight of which was eight or more tons, and gradually educate and accustom him to the great, strong and heavily ironed box, prepared and placed on the vehicle. Weeks of valuable time was consumed in preparation, and for a season it seemed as though all effort to get Jumbo to America must fail. At last, however, after sinking the wheels in the ground as described elsewhere, he entered the box, and was drawn by forty draught horses to the vessel, *Assyrian Monarch*, of the Monarch Line, and taken to New York. Of course, the cost of freight, duties and other expenses have been very great, but are nothing as compared with the value of this stupendous animal, which is without question the largest elephant in or out of captivity. These pages faithfully detail the sad regrets with which his English friends parted with him, and there can be no shadow of doubt but that the American public will esteem so distinguished and treasured a visitor with a regard fully as great. Tremendous and unparalleled is Jumbo in stature, and broad as the world is his fame. He is today the largest known animal that walks the earth.

Death and the Circus

CHARLES EDWIN PRICE

On April 10, 1882, Jumbo, still in his crate and fortified with a bottle of whiskey, was craned up from the foredeck hold of the steamship *Assyrian Monarch* and settled onto the dock at New York. With the assistance of sixteen horses and a few hundred men from the gathering crowd, Jumbo in his crate was placed on a trolley and hauled all the way to Madison Square Garden.

A show was scheduled for the next afternoon, so it was not long before the magnificent pachyderm was experiencing the smells of excited crowds, and the sounds and sights of brass bands and multiple performances taking place simultaneously in three rings wrapped by a race track forty feet wide and some four hundred yards long—and featuring in various combinations four hundred human performers, three hundred show horses, close to thirty elephants, and dozens of other animals, as well as numerous features and freaks, not to mention jugglers, clowns, acrobats, high-wire artists, and Mademoiselle Zazel being shot out of a cannon. This was the Greatest Show on Earth, and it lasted at Madison Square Garden for the next thirty days.[32]

After that, the Greatest Show began its touring season. The entire circus was transported in its own train (actually, three or four trains) with a hundred private carriages and flat cars, all hauled up and down the eastern United States and into Canada, stopping at towns and cities for a day or two, mostly. During an ordinary, nine- to ten-month touring season, there might be as many as a hundred overnight journeys, with the train arriving in a new town before dawn, after which the tents and equipment and supplies, the circus wagons, the animals—some caged, some not—and the people would be unloaded and assembled into an informal parade for the mile or two from the railroad siding to the performance field. Jumbo, as the great star of the Greatest Show, traveled in his own private railcar. But the nomadic life was grueling, and stresses were multiplied for the animals who were confined for long periods inside their railcar quarters.[33] Near the end of the first year of touring, in fact, Jumbo became seriously ill, with symptoms that included fatigue, poor appetite, and indigestion.

He died less than three years later, although not from indigestion or illness. On September 15, 1885, while returning from an evening performance and approaching the circus train on its railway siding for another overnight trip to another town, Jumbo was struck from behind by a freight train locomotive. The collision derailed both train and elephant, and one of them died as a result. The body was hastily protected from souvenir hunters by a fence and police guard, and a number of expert taxidermists from the Natural History Establishment of Rochester, New York, were soon on the scene to separate skin from bones from flesh. A crude autopsy, meanwhile, revealed a stomach full of foreign objects— stones, English copper pennies, coins of gold and silver, circus trinkets made from metal and glass, a police whistle, screws, rivets, a collection of keys, lead seals from railroad cars, and sections of wire once used to wrap bales of hay—which might account for the indigestion.[34] It took several months, but eventually 1,350 pounds of skin were realistically wrapped around a frame of wood and steel and set atop a wheeled trolley. During the next touring season, stuffed Jumbo was rolled around the Big Top, followed by a second trolley carrying his rearticulated skeleton, followed by his veiled and grieving elephant "widow." Other elephants from the Greatest Show on Earth waved black-edged bed sheets with their trunks and wiped their eyes at critical moments.[35]

For American circuses, Jumbo's case was different in detail but unexceptional in pattern. The unnatural conditions and intense stresses affecting elephants in American circuses during the nineteenth and twentieth centuries often produced comparably dire consequences.

By 1911, a quarter of a century after Jumbo's death, some thirty-two circuses were touring by rail across the United States. Some were small shows requiring as few as five railroad cars, but the largest of them required close to a hundred.[36] Elephants were an essential, almost defining part of the circus, and they were valued above all for their size: the bigger, the better. Bigger often meant more dangerous; indeed, one of the main attractions of a circus elephant was the frisson, the thrill of potential danger embodied in an enormous and seemingly alien beast. Since males grow to be twice the size of females, males were often the stars by virtue of their physical size—as Jumbo was—but males would also experience periods of musth, when testosterone levels spike dramatically and with dangerous consequences. Meanwhile, the technologies for controlling giant animals on the tour were still primitive: hooks and chains, ropes and wooden stakes pounded into the ground.[37]

So elephants periodically got out of control, went on a rampage, killed someone, and were then killed in response—often in secret. According to journalist Shana Alexander, author of *The Astonishing Elephant*, in America between the end of the Civil War and the start of World War I "dozens of circus bulls, and some difficult females as well, were variously shot, poisoned, stabbed, clubbed, garroted, electrocuted, drowned, and even hung by the neck until dead."[38] Circus owners often gave their female elephants male-sounding names, and because the differences between male and female elephants are not always obvious to the general public, their owners were able to disguise the fact that the males were simply being wiped out of the circus population; but by 1952, a census of circus elephants in the United States found only 6 males out of a total of 264 animals.[39]

Eliminating the males was a long process that began early. There was Columbus, a huge male who in 1841, in the town of Algiers, Louisiana, got into a ferocious battle with another male that was broken up by circus hands. An enraged Columbus then turned on his trainer, killing him and the horse he was riding before reaching his trunk over a fence to kill a second man and his team of mules. Columbus was shot. There was Chief, who in 1880 in Charlotte, North Carolina, killed his trainer; in 1889 he "had to be shot" as a dangerous animal. Mandarin obliterated a circus worker in 1881, while on tour in New Zealand. The circus owner planned to weigh down the elephant's crate and toss it and him overboard during the voyage back to the States; meanwhile Mandarin died from eating a box of kitchen matches, so his body was merely hauled out of the crate and pushed overboard. Tip, after attempting to crush to death his trainer, was killed at the end of the season. A second Tip, having obliterated three of his keepers, was "disposed of." In Racine, Washington, Prince killed the head elephant trainer; he was pitchforked to death by the other circus workers. In Peru, Indiana, a male known as Big Charlie wrapped his trunk around a keeper, picked him up, and summarily drowned him in a nearby river. Big Charlie was done in by a poisoned apple. Topsy, having killed a casual visitor who sadistically passed him a lighted cigarette, was electrocuted in 1903 at Luna Park of Coney Island in front of a paying audience. In Corisana, Texas, Black Diamond stabbed a housewife with a tusk and threw her husband over several railroad cars. He was executed by a twenty-man firing squad.[40]

Then there was Mary, an impressively large female and the star of the Sparks World Famous Shows circus. Badly handled by an inexperienced circus hand named Walter Eldridge one day in September 1916, she picked him up and threw him aside, then "placed her foot over his head and squashed it like a ripe melon."[41] She was executed the next day in a public hanging at the railroad yards of the town of

Erwin, Tennessee. How can anyone hang an elephant? Her story has been researched and dramatized by Charles Edwin Price in the short narrative *The Day They Hung the Elephant* (1992).

——— ——— ——— ——— ——— ——— ——— ———

As Charlie Sparks had predicted, Kingsport was a lucrative stop for the circus. The matinee crowd overflowed the bleachers, and the evening's performance promised to be even more profitable than the one in the afternoon. If the circus had not been scheduled to be in Erwin the next morning, Sparks would have considered remaining in Kingsport another day.

After the performance, the elephant trainers were scheduled to walk their charges a half mile up Center Street to a pond where the elephants would be able to drink their fill and wade around in the water to their hearts' content. Elephants, like children, love to splash in water. Eldridge was especially happy because he would finally get to ride Mary, the star of the show. The ponderous march to the watering hole was bound to attract gawkers; and, sure enough, at first sight of the elephants, a crowd of onlookers assembled along Center Street. Among them was nineteen-year-old William Coleman.

The sky was just beginning to overcast when the parade of five elephants began making their way up Center Street. Coleman remembered later that each of the beasts bore a rider, and each rider held an elephant stick to keep the elephants under control. Center Street, surfaced with hard-packed dirt, was a wide thoroughfare that sported a big drainage ditch running down the middle. Buildings were sparse along the route. Instead there were the open fields spotted with tents (which served as temporary dwellings for newly arrived Kingsport residents). In fact, on one stretch of the road, there was only one building—the ramshackle blacksmith shop of sixty-five-year-old Hench Cox.

Cox had not been to the circus matinee—he was too busy. A new town needed the services of a blacksmith badly, and he was far behind in his work. In fact, not even the commotion caused by the approaching elephants made him look up from his glowing forge.

The elephants, each with a trainer on her back, lumbered up Center Street, trunk to tail, with Mary leading the way. Several pigs, munching happily on a watermelon rind, scattered at their approach. The rind attracted the attention of Mary, who paused momentarily. Anxious to keep the animals moving, Eldridge prodded Mary with his elephant stick. Mary shook a little and snorted.

Once again, Mary reached for the rind with her trunk. The elephants behind her stopped, causing a roadblock. Some people in the crowd began laughing at the inability of the big elephant's trainer to make her obey. Anxious not to hold up the line any longer, and getting a little embarrassed at his lack of control, the impatient Eldridge whacked Mary sharply on the side of her head with his stick. Suddenly, the whites of Mary eyes flared as she wrapped her trunk around Eldridge's slim body; then she lifted him into the air. The crowd gasped.

Mary flung Eldridge through the side of a wooden soft drink stand. There was the sickening crunch of wood and human bones. Then the elephant calmly walked over to where Eldridge was lying, placed her foot over his head and squashed it like a ripe melon. Coleman said later he did not know whether Eldridge was already dead when Mary crushed his skull.

Women screamed, and onlookers scattered into the fields hoping to escape the mad elephant's wrath. The screams and shouting caused blacksmith Hench Cox to charge from his blacksmith shop. In his hand he brandished a 32-20 pistol. He saw Eldridge's mangled body on the ground, his head smashed to a bloody pulp. Big Mary was slowly backing away. The other elephants were loudly trumpeting, adding their noisy contribution to the screaming of onlookers. The trainers on the elephants behind Mary leaped to the ground. Some ran over to Eldridge, while others tried valiantly to keep their excited animals under control.

People were running in all directions trying to escape. Cox naturally assumed one of the elephants had gone berserk, and since Mary was the closest to Eldridge, he fired five times at her. Mary groaned and shook as the bullets struck, but they did not penetrate her tough hide.

Some people in the crowd, including William Coleman, stood their ground, though staying well outside the reach of Mary's trunk. A couple of roustabouts rushed to Mary's side and tried to calm her down. Seeing that the elephant was standing its ground and not attacking anyone else, the crowd began to reassemble. The sight of Eldridge's bloody body on the ground sickened some of them. Suddenly, they began chanting, "Kill the elephant. Let's kill him."

Shortly after 4 o'clock someone shouted, "Here come the elephants." The crowd craned their necks. A procession of five elephants, walking trunk to tail, ambled from the circus grounds onto Nolichucky Avenue, turning South on Tucker Street, then to Main. Miss Bondurant, from her vantage point atop the boxcar, noticed that the circus people followed the procession in double file—"some looking sad, some crying."

When the procession turned northwest onto Second Street, passing between Liberty Lumber on the right and Crystal Ice and Coal Company on the left, fireman Bud Jones thought something was wrong with the elephants. Mary was acting up—hesitating and bellowing at the top of her voice. The other elephants were joining in noisy accompaniment to Mary's trumpeting. Jones had an eerie feeling that *Mary knew exactly what was going to happen.*

From his vantage point atop the engine, W. B. Carr also noticed the elephant handlers were having difficulty keeping their charges together and moving. Mary continued to trumpet wildly. Then she stopped and squatted on the ground, and the handlers used the other elephants to get Mary up and going again. Carr told Talley that he couldn't help feeling sorry for the condemned elephant.

The lumbering procession crossed three spur tracks which ran across Second Street (the Jonesborough Road). The elephants noticed three sinkholes filled with water to the right and began moving off to the side. The handlers knew that if the herd made their way into those pools, it would be difficult getting them back out again. Fortunately for everyone concerned, the men were able to keep the procession in line, on the road, and moving steadily toward the railroad yard.

From Second Street, the procession turned onto an old dirt road that led to the railroad yards and crossed a muddy little tributary of North Indian Creek. Then they turned southeast toward the roundhouse, passing between the machine shop and powerhouse. The elephants passed waiting Derrick No. 1400, stopping just short of the roundhouse and turntable.

From his position on Ol' Fourteen Hundred, fireman Bud Jones, like Guard Banner, noticed the large number of people on the coal tipple. He hoped the tower could withstand the weight. He, too, estimated the crowd at about 3,000 people—more folks than actually lived in Erwin. Where on earth had they all come from? He was also amazed at the size of an elephant, now that he saw one up close.

Dave Bailey, on duty in the Clinchfield shops, was fixing a broken engine post with an acetylene torch when curiosity finally drew him outside to see what was going on. In fact, all Clinchfield employees, still at work, left their posts to watch the hanging. Well-scrubbed faces—male and female—peered from the windows of the brand new office building the Clinchfield had built the year before.

About five hundred feet farther down the track, railroad labor-ers and circus roustabouts had dug a hole for Mary's body after the hanging. Bud Jones said the hole was as "big as a barn."

Roustabouts quickly chained Mary's leg to the rail. Mary shook, swayed and trumpeted. Handlers began leading the other animals away so they could not see their companion being hung. Mary pan-icked and tried to pull loose, but she was securely fastened to the rail.

The crowd in the yard murmured expectantly. Puddles dotted the ground from the heavy rain. A sticky combination of mud, cinders, oil and coal dust clung to leather shoes like cement. An atmosphere of gloom shrouded the scene. There was a feeling of uncertainty as well as anticipation. Certainly the mood was not festive—rather, it was somber and subdued. Mary clearly sensed something was wrong. Her natural back-and-forth swaying gained a nervous edge. The extra adrenaline coursing through her body might give her enough strength to pull loose—then she would be impossible to control. There could be another killer rampage.

While they waited, rumors continued to circulate through the crowd that Mary had killed before, and if she got loose, she would run through the crowd and murder all of them. She was an outlaw rogue who lashed out suddenly, seized a hapless victim in her trunk, and dashed the body to the ground in a murderous rage. She had also recently acquired a nickname suitable to her reputation—"Murder-ous Mary." Had Mary really killed up to twenty men before, as some people were now saying? Certainly, she had killed a man in Kingsport. There were eyewitnesses. A full account had been published in the newspapers.

One local minister declared that Mary was demon-possessed. Exorcise the demon, he advised, and the animal would be fine. If Charlie Sparks had been a religious man, he would have given the minister leave to proceed with the exorcism.

Elephants are among the most intelligent and perceptive of mammals. From Mary's point of view, there were crowds of people standing around, but no tent. This was no performance in the sense that Mary understood it. Her familiar props were absent. No one had placed the huge ornate gold-braided blanket on her back or the fancy halter on her head. Mary also sensed a change in her handlers. Their usual firm patience in dealing with the elephant was absent, replaced with a strong sense of urgency. Mary had found herself in a railroad yard—a familiar spot for Mary. She might be prepared to work, but where were the gaily colored wagons to push up the steel incline, onto the flatcars?

As her companions trudged back to the circus grounds without her, Mary grew even more nervous. Elephants are social creatures, and the presence of others of their kind has a pacifying effect. Mary shook violently when she discovered she was abandoned. Little clouds of straw dust rose from her dark gray hide. Her ridiculously small eyes, whites now prevalent, watched in terror as her companions, walking trunk to tail, disappeared behind the powerhouse.

Suddenly, there was more trumpeting. From atop the boxcar, Ambrose could see the other elephants had stopped and were trying to reverse their direction. Handlers scurried around, trying to turn their charges back toward town. Obviously, the other elephants did not want to leave Mary. The handlers continued to work with the elephants and finally got them turned around again. According to Ambrose, no one dared to do anything to Mary until the other elephants were well on their way back to town and out of sight.

Ambling down Second Street, the returning elephants spied the water-filled sinkholes that lay just off the dirt road. This time there was little the handlers could do to prevent their charges from leaving the road. A moment later all four elephants were in the water, happily splashing away. It was fortunate that, from the sinkholes, the elephants could not see the deed being done to their leader.

Sam Harvey and Bud Jones were joined by sixteen-year-old Mont Lilly and two other derrick car crewmen, as well as a couple of burly roustabouts from the circus. A ⅞-inch chain dangled from the derrick boom. One of the roustabouts threw the end of the chain around Mary's neck and fitted the end through a steel ring. Mary started to bolt, and the other men scrambled to safety. Harvey dove into his control cab to get out of the way. The crowd of three thousand people waited breathlessly. Nothing happened for a long moment, then the nervous Sam Harvey threw the stick forward, and the winch began to squeal.

Slowly, the powerful derrick motor began reeling in the chain. In a moment, the chain began to tighten around Mary's neck, slipping taut through the ring. Mary's head was lifted. Her wind was being cut off. Her front feet left the ground. Mary struggled. In a moment, her hind legs lifted and began to wiggle back and forth, one leg at a time, as if in slow motion. Soon she dangled five or six feet above the muddy ground of the railroad yard.

Suddenly, there came a report like a rifle crack and ricochet, and Mary fell heavily on her rump with a sickening crunch. The cable had snapped. The killer elephant was loose. Immediately, everyone started to run. Bud Jones climbed the crane tower in a panic. Ten-year-old Lonnie Bailey took off and ran into a briar patch, scratching himself badly.

Blind Jim Coffey sensed the panic of the crowd; he had heard the thud when Mary hit the ground. He began running like everyone else. His sixth sense—that had served him so faithfully before—failed him, and he ran headlong into an equally panicked onlooker. Both were knocked to the ground and nearly trampled by the human stampede

running over them. They sprung to their feet, and the man yelled at
Coffey, "What's the matter, can't you see?" "Hell no," Coffey screamed
back. "I ain't seen a lick in twenty years. I just come down to watch
'em hang the elephant." Meanwhile, Mary sat on her haunches like a
big jack rabbit; she had broken her hip in the fall. The crowd needn't
have run, because Mary wasn't going anywhere. One of the roustabouts ran up her back like he was climbing a small hill and attached
a heavier chain. The crowd soon saw they had nothing to fear and
began slowly drifting back to the scene. Sam Harvey, once again, put
his winch into motion. Once again the chain tightened, and once
again Mary was slowly lifted into the air. Gravely injured by her fall,
she fought less this time around. The chain held, and a few minutes
later Mary fell limp. She was dead.

Cutting the Chain

CAROL BRADLEY

Carol Bradley's *Last Chain on Billie: How One Extraordinary Elephant Escaped the Big Top* (2014) is an exposé of the treatment of performing elephants that maintains as its dramatic focus the stories of two particular individuals. The first is Scott Blais, a young American man who during his early teens learned, while working at a Canadian safari park, what was expected of someone who wanted to work with elephants: the willingness and ability to exert maximum dominance over a formidable animal. The second is Billie, an Asian elephant who worked as part of a circus act for three decades but was never entirely dominated.

By the time he reached his twenties, Scott Blais had quit his job at the Canadian safari park, determined that he would never again club, hook, poke, whip, shock, or chain an elephant. With Carol Buckley, another former trainer who held similar opinions, Blais imagined creating an environment for retired zoo and circus elephants that would give them freedom, as much as possible, from human dominance, while being allowed to wander across open land and socialize normally with others of their kind. In 1995, Buckley and Blaise acquired 220 acres of former farmland in Tennessee and opened their Elephant Sanctuary.

Billie was born wild in India, then captured, isolated, immobilized, broken, and in 1966 placed in a large crate and shipped to America, where, at the age of four, she became an exhibit in a small, privately owned Massachusetts zoo. When she was ten, Billie was bought by a wealthy midwestern businessman and impresario, John Cuneo Jr., who owned Hawthorn Enterprises, which specialized in the buying, selling, leasing, and exhibiting of exotic animals trained to entertain the public. At Hawthorn, Billie learned to balance her entire weight on a single foot; to walk upright on her hind legs or her front legs; to rise up on two legs and form a line mount with other elephants; to dance, spin, pose, and so on. Soon, decorated cheerfully with red anklets and a big red headdress with a great golden star, she joined Cuneo's traveling troupe of performing pachyderms. For more than twenty years, this elephant's life consisted largely of huddling inside the back of a tractor-trailer semi as it rumbled from one performance venue to the next, then performing, waiting in chains between performances, and climbing up a ramp into the back of the truck for another rumbling journey to another show. In 1993, Billie and the other members of her group were trucked 2,000 miles to the West Coast, then transported in the hold of a ship for another 2,500 miles to Honolulu, Hawaii, where, according to a report filed by the US Department of Agriculture, she "attacked" her trainer and was reputed to be "difficult to handle."[42]

She was angry and dangerous, I can imagine, as were several other Hawthorn elephants. Some had merely knocked down trainers and handlers in a burst of pique. Some had done much worse. Joyce killed a handler who, one drunken evening, carelessly climbed onto her knee, wrapped himself up in her trunk, and put his hand inside her mouth.[43] Frieda killed two people, members of the public who were too familiar, too close, or too careless, and nearly killed a third. During another circus event in Honolulu, this time in 1994, Tyke trampled a trainer to death, shattered the leg of a circus promoter, chased a circus clown across a parking lot, and then rampaged down the street, crushing and pushing over cars until finally being stopped for good by eighty-seven bullets fired by members of the city police.

In 1996, Cuneo permanently retired some of his elephants, placing them inside a windowless metal and concrete barn in Richmond,
Illinois.[44] Billie and Frieda were among those retired to the barn, and,
since they were considered the most dangerous of the group, they
were kept together in their own twenty-by-twenty-foot stall. Workers
fixed an eight-foot chain to Billie's left front leg, so that as she walked,
she dragged the chain. This was for the safety of her handlers, who
could grab the end of the chain and tether her in place whenever
someone needed to enter the pen. Most of the other elephants spent
their days chained to the wall at the far end of the barn. But Cuneo's
elephant problems were only beginning. Soon it was discovered that
several of his animals were showing indications of tuberculosis; some
of his workers also tested positive for the disease. Other complaints
having to do with veterinary care, physical abuse, and allowing unsafe
contact with the public emerged over subsequent years, and on April
9, 2003, the federal government charged Hawthorn Enterprises with
forty-seven different violations of the Animal Welfare Act. As part of
the settlement of that case, Cuneo agreed to send his elephants to
places that could care for them humanely. Eight of them, including
Billie, were sent to the Elephant Sanctuary in Tennessee.

The following two excerpts from *Last Chain on Billie* describe Billie's
transportation from the Hawthorn barn to the Elephant Sanctuary (and
Scott Blais's initial removal of four feet from her eight-foot chain) in
2006. It would be five more years before Billie trusted people enough
to enable someone to remove the last four feet of her chain, and it
still required the assistance of two specialists in positive-reinforcement
"target training," Gail Laule and Margaret Whittaker. The second passage recounts that final event.

Two months later, on February 8, 2006, the morning of Billie's departure dawned sunny but freezing: the Midwest's notorious humidity made the thermometer's thirty-five degrees feel more like fifteen. Blais rolled up his bedding and shoved his belongings behind the passenger seat of the tractor trailer. He swept his fingers through his light brown hair and tried not to look as apprehensive as he felt. As soon as he could get the paperwork in order and load the elephants, he could loosen up. At least a bit.

Three times over the last couple of weeks the "Caravan to Freedom," as the media dubbed it, had made the six-hundred-mile trip from the Hawthorn exotic animal barn outside Chicago to the Elephant Sanctuary, delivering a pair of pachyderms each time. Minnie and Lottie were the first to arrive, followed by Queenie and Liz, then Debbie and Ronnie. The elephants seemed to acclimate quickly. They headed out to explore the two hundred acres set aside for them, conferring with one another in squeaks and low rumbles and caressing one another reassuringly. Minnie was especially adventuresome and energetic. She reveled in her freedom to wander: crossing the creek, pushing on trees, digging holes in the earth, and wallowing with abandon.

Even at this stage, John Cuneo was making things difficult. He'd made known his contempt for the Sanctuary—the notion that elephants needed saving from circus work struck him as ridiculous—and days earlier his employees had delayed Debbie and Ronnie's departure by several hours. Blais worried that, even with just two elephants left at Hawthorn, Cuneo could change his mind and throw a wrench into the plans. For weeks now, for harmony's sake, Blais had smiled and nodded and acted deferential around Cuneo and his employees. Blais was exhausted from all the tiptoeing around. He couldn't relax until all eight elephants were in the Sanctuary's possession.

He headed to the barn to check on Frieda and Billie. Two months had passed since Sue had died, and Blais had no idea what frame of mind he would find Billie in. Despite Frieda's violent past, Blais sized her up as less of a problem. She was thousands of pounds underweight and seemed perfectly compliant. The Hawthorn handlers considered Billie more volatile by far. Blais hadn't seen that side of her, but he knew it could surface at any time.

As much as Blais respected Billie's resiliency, he needed her to work with him, to board the back of the semi. He'd spent weeks mulling over the best way to coax her inside the trailer, and how best to keep her calm once the trip got under way. Billie and Frieda would be riding together and Blais decided to board Billie first. It might take three minutes or three hours; however long she needed was fine with him.

The saga of the Hawthorn elephants was big news in the Chicago area, and a CNN correspondent and camera crew were on hand to record the final road trip. A Hawthorn employee tried to block the camera's view of the barn. But to smooth the process, Wade Burck offered to remove himself from the scene. Blais had never seen Burck mistreat Billie, but he knew the trainer regarded the elephant as brooding and obstinate, and Burck's very presence in the barn caused Billie to bristle. "She hates me," Burck told Blais, and he wasn't exaggerating. Blais could tell from the way Billie planted herself so squarely, from the way she stiffened her trunk and hardened her gaze on Burck, that she really did despise him.

Once Burck moved out of the picture, the driver of the semi, Angie Lambert, backed the trailer to a ten-foot-wide gateway at one end of the barn and opened the rear doors. Billie's task seemed simple enough. From inside the barn, she needed to walk twenty-five feet to the ramp, step up the incline into the truck, and make her way to the front end. Getting on and off the circus trucks was never optional for elephants. They did it, and they did it on circus time. Handlers prodded them with bullhooks, yelled at them to keep moving and, if that didn't work, they whacked them—hard.

This time, Blais *asked* Billie to walk up the ramp. For once she had the freedom to think about her actions, to take her time. She had a choice in the matter. Blais had learned from experience to let go of any expectations and just let matters unfold as they would. This was Billie's day and he would heed to her schedule.

He stood back to see what Billie would do. Of all the elephants, she was the most distrustful, the most insecure. She might balk at the very thought of leaving the barn. She might even turn violent. One false step and Blais could have a disaster on his hands. But it occurred to him that Billie might also be curious about the trailer—that she might wonder what it was doing there and why she had been asked to step aboard. If the alternative was to remain in her Hawthorn pen, Blais thought, what animal wouldn't welcome the chance to escape by whatever means possible?

He called for Billie to come forward. She stepped out of the barn. The minute she laid eyes on the trailer she waved her trunk and began sniffing. She moved about slowly, lifting her trunk to feel the sides of the fences that lined either side of the walkway. An elephant's trunk is one of nature's most remarkable inventions. At three hundred pounds, it's powerful enough to hoist a person high in the air and dash him down on the ground, but also sensitive enough to grasp the tiniest of objects and relay all sorts of information back to its owner. Its pliable tip acts as both nose and hand. Billie was using her trunk to investigate everything about this new scenario.

Blais thought about how different her past road trips had been: tens of thousands of miles in dark chambers, anchored in place alongside a half-dozen other elephants. Blais wouldn't blame Billie if she never wanted to step foot inside a trailer again.

The elephant made her way to the trailer's back end. She brushed her trunk first on one side of the ramp and then the other. She glanced up inside the trailer as if to study its open and airy interior. She had to know that six other members of her herd had already left the barn. Surely she could detect their scents inside the semi.

"It's okay, Billie," Blais said to her. "You can do it."

Suddenly, without hesitating, she walked up the ramp and entered the semi. Slowly, deliberately, she continued down the length of the truck until she reached the front compartment. Then she turned and faced Blais as if to say, *Now what?*

Blais glanced at his watch. Fifteen minutes had passed. He'd been prepared to wait all day and the next to get Billie on board, and she'd finished the job in fifteen minutes. Maybe she sensed that this was no ordinary journey—that here was her chance to turn her back forever on her wretched past and start a new life. In spite of her fear, in spite of all her memories of traveling in chains, she'd demonstrated a remarkable measure of faith, a willingness to take a chance. Blais had asked her to climb on board, he'd told her she would be all right, and she had done as he had asked. It was as though she was saying back to him: *Okay, I'm going to trust you on that.*

The inside of the trailer was six feet wide and ten feet long, narrower than any compartment Billie had entered in quite some time. The semi was white and odorless and sparkled with cleanliness. Slatted vents the size of small windows allowed fresh air to flow in on either side. The truck radiated with heat, too. On this frigid February day, no blustery breezes sliced through corner cracks, no bone-chilling temperatures seeped up through worn-out wooden planks. Billie stood on cushioned, warm floors.

She had the front chamber all to herself. She watched as Blais now asked Frieda to step up the ramp. The smaller elephant did not hesitate. She climbed the ramp slowly and came to a stop in the middle of the rear compartment. A set of bars separated the two animals, nothing more. All that space for just two elephants.

Blais waited a few moments for the elephants to settle in. Then he worked his way up a protected aisleway on the right side of the trailer, knelt in front of Billie and took hold of the steel chain dangling from the bracelet clamped around her front left ankle. Quietly reassuring her, he slowly produced a pair of bolt cutters and clasped the tail part of the chain. He paused for a moment and then, heart pounding, he snapped that part of the chain in two. In one swift move he'd managed to lop off four of the eight feet. He could have snipped off more if he'd been able to get closer, but it was a start.

Billie watched him intently. She paced back and forth, up and down the ten-foot-long enclosure, testing her boundaries, what was left of her chain jangling beside her. She turned and once again glanced at Blais sideways with a look of concern, a look that said, *Don't push my limits.* But she seemed to also be registering the fact that no one had raised his voice to her or struck her from behind to force her inside. Blais had stood back and let Billie enter on her own time. Elephant time.

Blais ran down a final checklist with Hawthorn employees. It was midday when he climbed in the passenger side of the cab and Lambert pulled the tractor trailer out of the Hawthorn complex. The semi snaked its way on to Interstate 65, the CNN crew following behind. For the next seven hours the semi rumbled south through America's heartland, from Chicago to Indianapolis, stopping occasionally to feed and water the elephants.

"I don't think many folks at the truck stop in northern Indiana had any idea that when Blais was pulling a hose up to his truck, he was giving two elephants a nice big drink of water," CNN reporter Keith Oppenheim told viewers at one point.

Every couple of hundred miles Lambert pulled over so Blais could hop out and check on the elephants. The trailer had a large set of side doors, adjacent to Billie, that enabled Blais to work with them from the outside. He'd greet Billie first with soothing words and a treat: a couple of apples, a handful of peaches, a bunch of carrots, then make sure Frieda was doing well, too. If either animal had any fears about traveling, Blais wanted to quell them. But the elephants behaved calmly. To Blais's surprise, Billie seemed especially relaxed. She'd spent time in his company off and on for several weeks now and seemed to have sized him up as an ally, someone she could count on to treat her compassionately. Not once had she spied a bullhook in his hand.

A few miles north of the Kentucky/Tennessee border the driver pulled over for the night. Blais gave the elephants a final meal of the day of carrots, apples, bananas, and hay before he fell asleep a few feet away, propped against some hay bales at the front of the trailer. From time to time Billie and Frieda paced about. Shortly before seven o'clock the next morning Blais awoke, fed and watered the elephants, and then climbed into the cab next to Lambert for the last two hundred miles, bypassing Nashville and heading south. At noon they turned onto a gravel lane, negotiated a series of curves, then backed up. The truck moaned to a halt.

Word spread that the final truckload of elephants had arrived, and the Sanctuary staff gathered to greet the new inhabitants. Blais was too exhausted to celebrate the milestone. For two weeks he'd been on the road, back and forth to Illinois, nearly five thousand miles in all. He was bone tired. He wanted nothing more at that instant than a steamy shower and a single night's uninterrupted sleep in his own bed.

He stepped down from the cab, talked briefly with the staff, then strode to the back, unlatched the heavy metal doors and swung them open. The elephants stood silently. Blais beckoned to Frieda to disembark and she did so without any fuss. Moments later he returned for his final passenger. "Come on out, Billie," he said to her. "It's okay."

The elephant paused for a moment and flapped her ears. The tip of her trunk touched the inside wall of the trailer as if she debated whether to leave this clean, safe chamber. For the first time in her life she had ridden in a trailer with fresh air, plenty of room, and an abundance of hay and produce. But Blais was calling her to step out.

Carefully she moved backward, one enormous hind foot followed by the other. The four-foot-long chain hanging from her front leg clanged loudly against the metal floor as she made her way to the ramp.

"Very good. There you go," a second voice said.

Billie paused. She didn't recognize Carol Buckley's voice. "Take your time," Buckley said as Billie felt her way to the back of the truck. "What a *good girl.*"

The elephant edged back the way she'd done so many times in her past. A gust of cold air gripped the tuft of hair at the end of her long tail, her leathery backside and finally her broad, impassive face. Once she stepped off the ramp she turned toward the barn and immediately stepped indoors again. Caregivers crowded nearby, clapping quietly, beaming. They didn't want to startle her, but the significance of the moment left them gulping with emotion. Billie had finally escaped her awful past. After decades of confinement she was about to experience a return to nature she could never have imagined.

She didn't know that, of course. All she knew was that she was standing in a warm building bathed in light, a barn filled with the aroma of fresh hay and the familiar smells of her fellow elephants. And next to her was the man with the soothing voice who had brought her to this new place. He was speaking to her now. She might not have understood what he was telling her, but she recognized the sound of her name, and the tone of his voice was clear.

"Billie," he was saying to her. "Welcome home."

For five years Billie remained defiant, unable to forget her past. Her chain was a daily reminder of who she was and where she had come from. She'd worn it for so long she no longer seemed bothered by the noise it made. But to her caregivers, not to mention the Sanctuary's supporters, the chain was an ugly symbol of Billie's past. They were more than ready to see it go.

So Laule and Whittaker went to work. First Laule convinced Billie to present her foot when she was asked. Her caregivers lavished her with so much praise when she responded to their requests that Billie began to offer up her foot before she was even asked. Next, the consultants had two caregivers who worked on Billie, Richard Treat and Jennifer Hampton, familiarize her with the sight of the bolt cutters and how it felt when they rested against her chain. Three days a week for nearly three weeks, they practiced this routine.

One sunny May afternoon, Billie seemed ready. Whittaker stood just on the other side of the fence from Billie's trunk, talking reassuringly to her and poised to feed her a whole bucket of snacks. From the other side of the fence, Treat asked Billie to raise her foot. She did so willingly and rested it on one of the slats in the fence. Treat approached her gingerly, bolt cutters in hand. He reached through the slats and carefully snipped at one of the links in the chain. The link gave a bit. It was weathered and weak.

Whittaker and Treat waited a few moments, trying to gauge Billie's mood. It was crucial not to push her too far. Twice more Treat approached, knelt and made cuts with the bolt cutters. Billie swatted the fence with her tail, hard. Whittaker and Treat spoke to her gently to calm her down.

Each time Billie backed away from the fence, the caregivers' reassurances brought her back. It was almost as if Billie realized what they were attempting to do. The expression on her face softened and she stopped swinging at the fence. She lifted her foot again, this time higher than before, pushed up against the bars of the fence and rotated her ankle first one way and then another.

Half a dozen times Treat applied the bolt cutters to the chain on Billie's leg. After the sixth cut the chain remained in place, just barely. Treat set the bolt cutters down and, kneeling at Billie's foot, he reached out to disconnect the links. The instant he touched the chain it clattered to the ground.

If Billie understood the significance of the moment, she gave no sign. She glanced down at the worn chain, picked it up with her trunk, then dropped it and walked away. The small crowd that had gathered swallowed back emotion at the milestone, but the elephant had better things to do. She headed out to the sand pile to wake Frieda from her nap.

Abusing Captive Elephants in India

SHUBHOBROTO GHOSH

India today harbors the most significant population of Asian elephants in the world. A few thousand of them have been trained to serve humans in logging, construction, transportation, and tourism; the remaining wild herd is primarily contained within an area of 65,000 square kilometers distributed within twenty-nine forest reserves. With ever-expanding human numbers, however, elephants are continuing to decline across India, as they are everywhere else in their former range in South and Southeast Asia. As their forests are degraded by human activity, these animals move increasingly into areas of human habitation, where they are killed by poisoning, poaching, electrocution at high-power lines, and collisions with trains. One estimate suggests that an elephant is killed by human action once every four days in India, on average, while the total elephant population of that nation has dropped by 10 percent during the past five years.[45]

In the following essay, journalist Shubhobroto Ghosh reports on another aspect of that ongoing tragedy: the abuse of captive elephants used for shows and rides in the Indian tourist industry. Born in Kolkata, West Bengal, and educated in England and India, Ghosh has covered the science and wildlife beat for the *Telegraph* and worked for the Wildlife Trust of India. For traffic at WWF India, he investigated the live elephant trade at Sonepur in Bihar, while more recently he has focused on the plight of India's captive elephants as Wildlife Projects Manager for World Animal Protection in India. Ghosh's work has been published in the *Hindu*, the *New York Times*, the *Statesman*, the *Telegraph*, and other publications. He is the author of *Dreaming in Calcutta and Channel Islands* and has contributed to several other books, including *The Jane Effect*, a biographical tribute to Jane Goodall, edited by Dale Peterson and Marc Bekoff.

It is 5 a.m. at Haathi Gaon in Jaipur. It is dark. Nearly everybody is still asleep, but Rani must stay awake. She is standing in her urine and faeces, her movements restricted by chains and ropes tied to her feet. She can lift her rear legs, one at a time, when she is uncomfortable and does not like to stand in her own mess. Despite what her owner claims, Rani spends her nights chained and unable to lie down or rest completely, and this amounts to torture. In the wild, she would take short naps standing up, but she would also lie down to sleep for a few hours.

At around 7 a.m., it is time to go. Rani is ordered back to her stall. She has been trained to obey more than thirty-five commands: to move, up, down, sit, lift feet, turn around, pose for a photograph, and so on. She does not resist her mahout. She knows what she must do and what can happen if she wavers, so she obeys him and moves slowly. She kneels to have her howdah, an elephant saddle weighing about 50 kilos, lifted on her back. She waits patiently until it is fitted on tightly.

Rani then leaves her stable, her mahout riding on her neck, to make the arduous journey along the road to Amer Fort. It is early, but the Indian sun is already very strong. Her commute is dangerous and risky. The road is chaotic and noisy, and the tarmac surface is hot. The foot pads of an elephant are not made for paved roads, and so Rani's feet are all cracked on the bottoms and have developed sores. Of course, elephants have been killed on this road by careless drivers, but Rani has no choice. This is what she must do every day. She is used to it now, but she should not be because she is not meant to be there in the first place. Elephants are not native to Rajasthan, which is too hot, too dry, too dusty. They belong in the forests of greener states.

At around 8 a.m., after a stressful journey, Rani arrives at the lower levels of Amer Fort, where she meets her counterparts, dozens of working elephants lined up and ready to take tourists up the steep slope of the fort. Without time to rest and recover from the journey, she is forced to take her first customers of the day straightaway, a maximum of two people per ride. It takes about half an hour of slow and painful uphill-walking. The road is made of jagged stones, and many elephants go up and down at the same time. At the top, the tourists get off, and then, without a moment's pause, Rani is walked back down. She is made to go up and down that route four times, without rest or access to drinking water. She cannot stop or will likely receive a beating from her mahout, who still uses the bull hook, which is officially banned from use. Sick elephants are also forced to work.

A ride costs about $17 a person, so an elephant will earn some $130 a day. This income supports not just the mahout and his family but also the elephant's owner and his family. There is an Elephant Owners' Association to look after the interests of the owners and mahouts, and they are solidly in favour of the elephant rides, citing the need to earn a livelihood as their reason for perpetuating the cruelty to the elephants of Jaipur. According to Brigitte Kornetsky's remarkable film on the Jaipur elephants, *Where the Elephant Sleeps*, the owners are also averse to modern medical treatment for their elephants.

At around midday, when she has finished working at Amer Fort, Rani must return to the elephant village. She is then forced to stand in her stall with the howdah still roped to her back, waiting to take more foreign tourists for rides around the village during the afternoon. These rides are not regulated, so mahouts will do as many as they can, disregarding the welfare of elephants.

Although surrounded by people for most of the day, Rani is lonely. She has almost no interaction with other elephants, which is important for her psychological well-being, since elephants are gregarious animals with strong social bonds. Leading a solitary life is thus additionally stressful for an animal who in the wild would be part of a close-knit family. Only when the sun goes down is Rani finally done for the day. Tired, sore, and hungry, she is fed more sugarcane straw and is chained up for another long, wakeful night. This routine will carry on until she drops dead from disease, overwork or accidental injury.

Rani's plight was observed firsthand by an animal welfare investigator from World Animal Protection in London, Marie Chambers, who visited her in Jaipur not long ago. But she is one of well over a hundred elephants at Amer Fort in Jaipur, Rajasthan, who suffer chronic stress, heat and physical abuse. Indeed, abuse is rife among the elephants in Jaipur, who endure repeated beatings, inadequate diet and long hours of work. Regular research, including recent studies conducted by a variety of wildlife protection and animal welfare organizations, shows that the current status of captive elephants at Jaipur is morally unacceptable—and yet newly acquired elephants continue to arrive there, with illegal transfers from Bihar to Rajasthan done with forged certificates.

India has around 20,000 wild elephants and some 3,000 captive ones, many of whom are used for joyrides in places like Amer Fort in Rajasthan and for displays in Kerala. Elephants are also used in national parks and forests where tourists ride them to see wildlife. Tourists, especially those from abroad, spend exorbitant amounts of money to ride these majestic animals, and why not? For these tourists, riding an elephant is a dream come true, the high point of a fantasy that reinforces India's stereotype as a country of kings and mendicants, snake charmers, tigers and elephants. Yet most people who ride the elephants do not realize that these sentient and social creatures are brutally exploited in order to provide humans with a few minutes of fun.

Elephants are revered in Indian history and culture, featured in classical poems, epics, and novels as the living embodiment of majesty and nobility. Indeed, one of the most powerful images of India is that of Ganesha, the elephant-headed god. It is ironic, then, that this same culture has tolerated the brutal capture and taming of these gentle and noble creatures for thousands of years, and seen their exploitation as beasts of burden and transport and war—and, today, as sad captives forced to provide a happy few minutes for tourists. The Global Elephant Charter, which has been signed and endorsed by many eminent field biologists, scientists, and conservationists, states that "elephants are complex, self-aware individuals possessing distinct histories, personalities and interests [who therefore] are capable of physical and mental suffering." Banning elephant rides may one day bring some significant succor to the revered and yet much-abused living representatives of Ganesha, the great elephant god of India.

SOCIAL
AND SEXUAL
ELEPHANTS

Individuals

IAIN AND ORIA DOUGLAS-HAMILTON

In 1965, a young zoology student from Oxford University named Iain Douglas-Hamilton arrived in Tanzania, East Africa, hoping to study lions in the Serengeti. But John Owen, director of National Parks for Tanzania, told him about a pressing need for someone to learn about the elephants in the Lake Manyara National Park.

The park, which consisted of a shallow alkaline lake and a thin strip of land surrounding it on the northern half, happened to be one of the smallest parks in Tanzania, but it still contained plenty of animals—hippos, rhinos, buffalos, leopards, lions, and elephants. Manyara's most compelling tourist attractions in those days were the tree-lounging lions. Lions liked to hang out in the branches of the thorny *Acacia tortilis* trees. Indeed, particular lions had claimed particular trees as their favorite resting spots. Meanwhile, elephants were stripping the bark off Manyara's acacia trees and thus killing them at what seemed like at a rapid rate. No one knew how rapid the rate was, however, or whether this process was a natural thing, part of an ecologically stable cycle, or an unnatural one caused by too many elephants. Was it a problem? Something to be controlled?

No one knew much about the elephants either, although it seemed as if the park had plenty of them. It happened that the park ecosystem was rapidly being isolated by farms and settlements, so perhaps the elephants were being driven off a much larger range and concentrated into the smaller area. A casual survey done from a small plane had counted some 420 elephants altogether at Manyara, which was a very large number for such a comparatively small park. If true, it meant a density of around twelve elephants per square mile. That would be an extraordinary concentration of elephants, far beyond densities documented elsewhere in Africa. So were there too many elephants at Manyara? Were they growing in numbers or stable? Were they leaving the park seasonally? Were they killing the acacia trees beyond any capacity for regeneration? Iain Douglas-Hamilton was given a one-year research grant from the Royal Society in England for a study titled "The Feeding Habits of Elephants and the Way in Which They Modify the Vegetation," and the Tanzanian National Parks provided him with an assistant, a beat-up Land Rover, a prefabricated tin hut, and permission to live anywhere inside the park as long as he kept out of sight of the tourists. Although his original research grant would support him for only a year, the young zoologist ultimately spent four and a half years among the elephants of Manyara.

The urgency of his work was clear, since the question of how many elephants could be supported by the park was part of a larger debate between those who wanted to regulate elephant numbers by shooting them and those who wanted to respect something closer to a natural process. In order to produce a sound scientific opinion about the matter, Douglas-Hamilton would have to count trees and monitor them, and he would need to understand the dynamics underlying the elephant population: births and deaths and rates of fertility. But in order to gain that understanding, as he recalled in his 1975 memoir, *Among the Elephants*, he first needed to acquire the "ability to recognize large numbers of individual elephants with no more difficulty than an equivalent number of men." Knowing a large number of elephants as individuals would be the route to understanding their behavior and lives, he believed—yet no one had ever done it before.

But how could anyone recognize individual elephants? At first glance, they seem fundamentally indistinguishable: just solid, stolid, and gray. Perhaps elephants recognize each other individually through smell or sound, but for a visually oriented human primate, what was the logical way to mark and recall individuals? Douglas-Hamilton decided that he would photograph all the elephants and try to compile a catalogue of photographs that would become an elephant *Who's Who*. He was surprised at how difficult it was even to get decent photographs. Most of the time, the animals were obscured within "a dark cloak of vegetation." It was hard to find them out in the open, and even when he did, they would usually be inconveniently clustered together. Often only their backs were visible, with various other fragments—such as a flapping piece of ear or a waving trunk—periodically appearing and disappearing.

In time, Douglas-Hamilton began to solve the puzzle of individuality, and eventually he was able to recognize virtually every one of the elephants coming and going in Lake Manyara National Park. And as he developed his sense of their individuality in appearance, he also began to appreciate that the elephants were individuals in personality and character. He gave them names that often recalled distinctive features in visual appearance—and often in personality and character. In the following excerpt from his memoir (cowritten with his wife, Oria), Douglas-Hamilton tells of the process he went through in learning to recognize individuals and of his first encounter with the strong-willed matriarch he named Boadicea.

——— —— —— —— —— —— —— ——

Learning to remember an individual became like a geography lesson, in which the shape of a country's borders had to be memorized. Often an ear would be almost smooth, with only one or two small nicks, but the shape of the nick, whether it had straight or curved sides, its depth and position on the ear, provided useful material. Some nicks looked as if they had resulted from the ear catching on a thorn, others as if they had been deftly cut by a tailor's scissors in neat straight lines. Certain elephants had ears with as many holes along the edge as a Dutch coast line plastered with bomb craters along its dykes. The cause of these holes I never discovered, but I suppose it must be due to some internal physiological process, the result of which gave their ears a decaying appearance.

There were also the particularly large rents that came from the center of the ear. Often in young animals these looked recent and some were still bleeding. I later discovered that the cause was the intolerance of older animals, and the instrument the sharp tusks of some old cow who had jabbed the youngster through the ear, either leaving a hole or ripping the ear from the middle outwards in a long tear.

In time I found that the details changed slowly, and that ragged ears full of holes changed more rapidly than those with clean straight-edged cuts. Very young animals and a minority of adults had almost completely smooth ears; in such cases I had to record and memorize minute holes, no bigger than a piece of confetti, which would reveal themselves only under a close scrutiny. These tiny holes changed hardly at all. If the ear was clean of mud, it was possible to discover a fine network of blood vessels protruding from beneath the thin ear skin and visible as a delicate tracery of ridges. Photographing these could only be done when the light was slanting at the right angle, but if they could be recorded I was sure that they would change very little over the years, whereas over a long period an elephant might obtain sufficient new cuts on his ears and chips off his tusks to become unrecognizable.

Tusk chips and breaks and broken edges are gradually worn smooth. An elephant's tusks continue to grow all through its life. The rate at which they wear down is illustrated by the calculation made by Dr. Richard Laws that if they did not break during a life-span of sixty years they would reach a length of sixteen feet in the female and twenty feet in the bull. These breakages together with alignment give the tusks great variation. One is generally used as a master tusk and is worn down at a faster rate than the other, and therefore it is usually shorter and more rounded at the point. Very often the master tusk acquires a groove near the tip where the elephant habitually pulls grass over the same place.

One warm evening when I was on the beach I met a large concourse of elephants in the open. Here was a chance to get some frontal close-ups outlined against the sky. They were all standing, some eating the short spiky grass, others drinking from small holes which they had scooped out, the youngsters chasing each other and engaging in furious mock battles. I saw three one-tusk cows, several with moderate sized tusks and two large and beautiful cows with long curving tusks that swept together in a great gleaming bow. The older of these had sunken hollow cheeks. She stood peacefully, kicking up little morsels of grass with her front toenails, but the other cow cocked up her head, held it slightly on one side, and looked at me intently.

I was about two hundred yards away and clearly the presence of the car was worrying her. None of the other elephants paid any attention, so I took photographs of this panorama of elephants silhouetted like a frieze against the Lake. But as I continued to watch, the larger and more nervous cow began to pace back and forth. Then she stopped, and shook her head rapidly so that her ears flapped like stiff blankets being shaken and dust flew up into the air. Gradually she worked herself up to fresh demonstrations weaving to and fro in front of the group, always looking in my direction. Seeing this a few of the other cows became disturbed and moved in to stand behind her, spreading their ears and twirling their trunks; their calves were by their sides.

I made several counts and came up with the figure of forty. The large cow slowly edged her way in my direction, and the others followed in her wake. Their intentions were plainly aggressive. They reminded me of some massive biblical phalanx with a champion standing out in front of the army. The ground was flat, so knowing that I could leave my escape to the last minute and still be sure of getting away, I decided to test their intentions.

When the great cow came to within forty paces of me she stopped and drew herself up to her full height; her fellow matriarchs fell in behind her. I switched on the engine and she broke into a lumbering charge, her trunk rolled tightly up beneath her tusks like a coiled spring. I let her come to within ten yards, to see if she would stop, but she kept on at full speed so I let out the clutch and raced away keeping just in front of her, encouraging her to think she had a chance of getting me. There seemed little doubt that she was in deadly earnest. After some fifty yards she stopped, stood tall again, and emitted a resounding trumpet. Her posture was perfect for an identification picture, and shaking slightly I took it. All forty elephants had closed up behind her in a tight-knit mob.

Despite the strict training I had been given at Oxford not to give human interpretations to animal behavior, it was impossible not to anthropomorphize. This cow looked such a fine warrior queen that I named her Boadicea, after the ancient British chieftainess who, "earnest, rugged and terrible," had defied the invincible oppression of the Romans, fighting for her people to the bitter end.

Families

IAIN AND ORIA DOUGLAS-HAMILTON

Once he had begun to identify individuals, Iain Douglas-Hamilton re-alized he was in a position to sort out the Manyara elephants' social structure. Much of the traditional lore of hunters suggested that elephants were led by old males, the impressive "bulls." An earlier scientist had in 1957 hypothesized alternatively that the primary social unit for elephants might actually consist of closely related females and their young. But were those groups actually stable over time? Were they functioning as consistent and coordinated social units?

Meanwhile, Douglas-Hamilton was seeing herds of elephants that included up to a hundred individuals, although it was clear to him that such large groups were not stable over time. Eventually he began to believe that the primary social unit for the Manyara elephants was indeed a family unit: one led by a single older female, the matriarch, but including other closely related adult females—sisters, mainly—and their offspring. Nevertheless, recognizing the family unit was complicated by the observation that the Manyara elephants were often gathering periodically into those larger groups, the question then became: Were the larger groups organized in any way or were they simply random gatherings of the smaller family units? Because he had learned to recognize individuals, Douglas-Hamilton was eventually able to answer that question satisfactorily. The larger groups were not random. Family units preferentially gathered with a few favored other family units. Douglas-Hamilton believed that the larger preferential associations were based on extended lines of kinship, and so he called them *kinship groups.*

Comprising almost fifty individuals altogether, the Boadicea kinship group (that is, Boadicea's family combining with those of two other matriarchs, Leonora and Jezebel) was the largest at Manyara and the most frequently observed. From 1966 to 1970, Douglas-Hamilton observed that particular kinship group on 314 separate occasions, and he came to appreciate the distinctive personalities of several individuals in the group. Virgo, a small female with only one tusk, was "the tamest, gentlest and most curious elephant" of them all—while Leonora, the second matriarch in the group, never seemed to be disturbed by his presence and never aggressively displayed or charged. Rather, she remained predictably "stately and unruffled by even the closest approach of a vehicle."

But Boadicea herself had, along with the biggest tusks of any female, among "the fiercest threat displays," and her charges were "easily the most impressive." Nervous and sensitive, Boadicea was always ready to threaten and quick to charge. To Douglas-Hamilton, the charges were, at first, perfectly intimidating. But he began to notice that she hesitated before charging, showing signs of uncertainty by a tentative fidgeting of the trunk and the slow swinging back and forth of a front foot. And when she did charge, he saw that about ten paces before she reached his vehicle she would invariably "skid to a halt in a cloud of dust." Boadicea's charges, he decided, were mainly bluff, and over time he concluded that for all the matriarch's sound and fury, her charges signified little.

By the middle of 1966, Douglas-Hamilton had become familiar with nearly all the elephants who regularly appeared in the northern part of the park, including the Boadicea kin group, and he presumed that he was close to his goal of learning to recognize every elephant at Manyara. One morning while exploring the more rugged and dense southern end of the park, in an area near a waterfall of the Endabash River (which flowed into Lake Manyara), he came upon a group of elephants he'd never seen before. He saw four large females quietly grazing in an area of tall grass, and since a high wind masked the sound of his engine, he was able to approach within viewing distance before they noticed him. He cut the engine, hoisted himself up through the Land Rover's roof hatch, and readied himself to observe and photograph while sitting on the roof. Suddenly their four heads turned in unison while their ears raised "like hostile radar scanners about to launch a missile." One of the four tossed her head crossly, and then, with no further warning, they attacked. Used to the bluff charges of Boadicea, the young researcher stayed where he was, expecting them to stop well short of the vehicle. They did not. At the last second, however, they swerved away, while one of them shattered a dead branch with her tusks and then, looming above him, "let out a strangely savage and piercing trumpet that sounded as if she were using all her breath and emotion to expel it." These new elephants, he thought, were entirely different from any he had previously encountered, and he recognized them to be "totally hostile and irreconcilable to man." Why they stopped their attack that time, he would never know. He named them the Torone Sisters, in honor of "a shrill queen of Greek mythology."

In short, not only did individual elephants show marked differences in personality and behavior, it seemed, but so did the family groups. That first frightening encounter with the Torone Sisters was followed by later ones that became progressively more serious. Hugh Lamphrey, the Tanzanian Parks official who supervised the research at Manyara, had insisted that Douglas-Hamilton be able to recognize all the elephants in the park. So despite his own concerns about the dangers of accidentally coming across those hostile elephants, or the occasional rhino, he continued to explore the Endabash thickets. The worst attack by the Torone Sisters happened during a drive one day when Douglas-Hamilton was accompanied by a park ranger (Mhoja Burengo), a park worker (Simeon), and a friend (Katie Newlin). The vehicle was destroyed.

—— —— —— —— —— —— —— —— ——

W hen I returned to Manyara the rains had begun in earnest. Gone were those pleasant mornings with the sun streaming across the breakfast table. Slate grey clouds lowered for more than half the day and the Ndala River ran red and angry. The elephants were around in large numbers, but it did not rain enough to cause them to retreat up into the hills to get away from the sodden woodlands as they had done in other years.

As soon as the walls of my house were up and a roof was complete I moved in, glad to evacuate my dripping tent. The walls were painted white on the inside, and geckos soon made their homes in the eaves and hunted flies there. A desk, a table, some chairs and a bed were sufficient furniture.

I organized my patrols to take me to every part of the Park, and day after day increased my score of known elephants, at the same time experimenting with ways of measuring their height so that I could know their ages. One of the first questions I had to answer was what constitutes a stable elephant group. Once I knew this I would be able to study how these basic building blocks of elephant society interacted under the crowded conditions of Manyara, and how social behavior in turn affected numbers. As early as 1961, an American scientist, Irven Buss, had suggested that elephants assembled in family units of closely related cows and their offspring, but no one had yet been able to show that these groups were stable.

The groups I encountered usually had several cows with their young, which might be tightly coordinated, or which might be loosely strung out and overlapping with another group, so that it was impossible to see where one group ended and the next began. The largest groups were of the order of eighty to a hundred elephants, with bulls strung out on the outskirts and cows and young calves tightly clustering in the center, but these large herds never stayed together for more than a few hours before breaking up into many smaller groups.

With these first fragmentary observations I began to wonder if elephants did have any sort of social organization, or if individuals roamed at will and just joined up with whomsoever they met.

Gradually, I discovered the pattern of their movements and the best times and places to watch them. Starting early every morning I would drive along the foot of the escarpment looking up at the steep slopes, where the elephants usually spent their nights feeding on the variety of plant life found there and nowhere else.

With the first rays of sunlight warming their flanks I could usually find some group of elephants rocking their way down the escarpment to gain the shelter of the *Acacia tortilis* woodlands before the sun grew too hot or, if it were cloudy, they might stay until the weather cleared at about midday. One elephant after another would move in and out of view along the green tunnels carved as a network up the slopes. Steepness was no barrier. They tested uncertain corners with their trunks, gently probing for loose ground and slowly transferring weight in a smooth motion from one foothold to the next. The intermittent visibility meant that I had to wait for the elephants to clear the hillside before I could recognize more than a handful of them, but I could sense their group coordination merely by listening to them calling to one another in deep rumbles which were answered up and down the length of the hillside, each little party apparently keeping in touch with the others in this way.

Below, in the woodlands, there were smooth dusty patches under the trees, and here the elephants paused and rested when it was hot.

They would arrive one by one, each cow bringing her calves with her. It was in one of these dust bowls that I found Boadicea again about a month after I first saw her on the beach. From the uppermost branches of a nearby tree I was able to watch her enjoying a peaceful mid-morning siesta.

Standing on her right was another large cow, with fat convergent tips to her tusks, and two calves at heel; one had a wart on his head and the other had a wart on his trunk. The next largest cow also had convergent tusks, but these were thin and sharply pointed; she had a distinctive cut out of her left ear the shape of the Gulf of Suez. Another, whom I named Right Hook, had a tusk that curved sharply inwards and there was a small one-tusker female, Virgo. All had been on the beach together, and now I quietly took their pictures for a second time as they stood with eyes half closed and trunks hanging immobile or loosely draped over a tusk. I could only see twenty-two elephants. On the beach there had been forty.

Noises of a second party came from nearby. I crawled down the branches of my tree and found a better perch in the fork of two boughs, underneath the canopy. This was well within the reach of Boadicea's trunk if she cared to take a few paces in my direction, but she was unaware of my presence. From here I could get a good view of all directions.

Standing under a tree, not more than a hundred yards away, was another fine matriarch with long white tusks that curved gracefully inwards, longer but very similar in shape to Boadicea's. Her ears had relatively smooth edges, but her temples were sunken. She was the most beautiful elephant I had seen, and I remembered quite clearly that she too had been on the beach, mixed in with all the others. I named her Leonora. There was now a distinct unit of her own round her. I looked at them carefully to compare them with the pictures I had taken before. Sure enough, there was the same cow whom I named Slender Tusks next to her, and close by a barely mature female with a big V-nick out of her left ear. It took me most of the morning to count her group of nine, but for once luck was with me. The wind held steady, blowing from the south-east across the Lake, and taking my time I was able to maneuver round Boadicea without disturbing her or any of the other elephants.

Under another tree two hundred yards from Boadicea I discovered a third distinct unit. In it were the two one-tuskers who had been on the beach. I counted up this group with growing excitement. The numbers came out right; there were exactly nine. This made forty elephants altogether, and included all the distinctive females I had earlier seen together. The only difference was that now they were arranged in three smaller groups. My puzzle was beginning to work out. The groups were stable.

The formations I have just described proved to be typical. At times Boadicea would be in a large herd of forty, but more often there were three separate smaller units led respectively by herself, old Leonora, and the largest of the one-tuskers, whom I named Jezebel. Each smaller unit remained stable in itself, bound to family ties. These three family units were usually within a few hundred yards of each other, and I believe that together they belonged to one large kinship group, every member being inter-related.

During the next few months of 1966 I established that a similar family unit social organization applied to all the other cow-calf groups in the Park, of which there were at least forty-eight. The average size of the family units was ten elephants, and most of these belonged as well to larger kinship groups. Family units who were members of a kinship group might split up for a few days and go to opposite ends of the Park, but they would always join later and continue to keep company.

This discovery came as a great surprise to me, because up to this time, although there had been a hypothesis that the cow-calf groups might be stable family units, it had never been suggested that larger herds were anything but aggregations of family units joining and leaving at random. My observations provided the first proof of family unit stability and showed that family ties were far wider and more lasting than had been thought.

When one day I confessed to Hugh Lamprey that I didn't know all the elephants in the Park he replied, "Well Iain, you had better get to know them. You have chosen your method and you had better make it work."

After this, however much I disliked the idea, there was going to be no alternative to penetrating the thickets and meeting the Endabash elephants on their own ground. I did force myself to walk back to the exact spot where I had met the rhinos, this time with Mhoja and the big gun, but there was a group of Endabash elephants in possession, who set up a loud commotion as soon as our scent was carried to them on the breeze, and when we tried to work our way round them, we bumped into another rhino. Luckily he crashed off in the opposite direction, probably because that was the way he happened to be pointing at the time.

Down some of the trails it was possible to force a passage with the Land-Rover until it brought one to the river where they had come to drink. Here I waited for thirsty elephants to come, but often the wind changed; I was then upwind and nothing would induce an Endabash elephant to come anywhere near the water because of the smell of the car or of man. On other days I simply chose the wrong spot for that day's drinking.

It was therefore with a feeling of relief that one day I encountered Queen Victoria, who had come down to this area attracted by the sweet ripe *Balanites aegyptiaca* fruits, known as desert dates, that were then in full fruit. I had a friend, Katie Newlin, a member of the American Peace Corps, staying for the weekend. She was delighted to see elephants at such close quarters. The whole family was following a bull who went from one tree to another, putting his trunk up each and shaking it until the fruit tumbled down in a shower. Victoria's family would then rush up and join in the feast, which the bull tolerated in a good-natured way. Only the elephants' backs were visible above the bush as my stripped-down Land-Rover edged towards them. Mary, the other large old cow in this family, standing in a small clearing, barely looked up from her food, but she shook her head in mild irritation when a thicket splintered as I drove over it. I was glad to see them, as they had been missing for over a month.

Mhoja, standing in the back, spotted another group, composed of strangers, which he could see through the foliage. I drove towards them crushing branches in the way. In front of me a young female with a small calf whom I could not recognize ran off in alarm, behind a gardenia bush. Seconds later a huge, bow-tusked female came headlong round the gardenia and without uttering a sound, nor pausing in her stride, plunged her tusks up to the gums into the body of my Land-Rover. Mhoja, and a temporary laborer Simeon, who had remained standing in the back just behind the cab, saw the tusks appear beneath their feet and, with the huge shape looming over them, jumped out of the car and vanished into the bush.

The first shock threw the car half round. The elephant pulled her tusks out and thrust them in again.

"Don't get out of the car," I shouted to Katie. She lay down on the floor.

Now more elephants with babies in the forefront burst out of the bush on the right and joined in the attack. A three-year-old calf butted the mudguard and then stood back bewildered. Fortunately, because of our relative sizes, only three elephants could concentrate on the Land-Rover at one time, but it was enough. It felt half-way between being a rugger ball in a scrum, and a dinghy overtaken by three contradictory tidal waves. The car teetered on the point of balance but just missed overturning. Tusks were thrust in and withdrawn with great vigor. Loud and continuous trumpeting rent the air, together with that fatal sound of tearing metal. However, I was not thinking just then of what John Owen would say about my new Park's Land-Rover, because an enormous brown eye embedded in gnarled skin with long eyelashes materialized through the upper door. This belonged to a cow who was using the weight of her head to force down the roof of the cab. The cab cracked and gave, but then held firm, while her tusks ripped sideways across the door. I could have touched the eye with my fingers. To my relief it disappeared without apparently perceiving that it had been looking at the undamaged brain of this metal animal. I could imagine her picking off our heads just like bananas off a bunch.

A huge latecomer with as much zeal as the rest put together now came into contact with the front. One wing folded up like paper and a tusk went through the radiator. She stabbed again and wrenched her embedded tusks upwards like a demented fork lift. Then digging her tusks in again she charged, and the Land-Rover was carried backwards at high speed for thirty-five yards until it squashed up against an ant heap surmounted by a small tree.

They left us now at rest, adorning the ant heap, and retired thirty yards, where they formed a tight circle and, after a few excited trumpets and growls, dissolved into the bush, with streaks of green paint on their tusks.

My Peace Corps friend picked herself up and dusted her blouse. She was a little shaken, but unharmed, and accepted the situation with sang-froid. My first awful thought was what had happened to Mhoja. The only response to my shouts and whistles was sullen rumblings. The car looked a write-off, but I pressed the starter and to my amazement it worked. One tire was punctured, a tusk had passed clean through it, and twisted metal scraped against one of the front wheels. However, it was possible to lever the metal away, and we limped off on the flat tire to begin our morbid search. Elephants were all around but their destructive rage appeared to be sated.

At the first point of impact we stopped. There was no bloody smear on the ground here nor even a chip of ivory. I stood on the Land-Rover and shouted again. A faint mocking echo came back from the woodlands, or was it a shout? I drove on deeper into the bush and tried again. This time there was a definite answer. Eventually, after about a mile, Mhoja's green uniform materialized out of the twilight.

We were both relieved to see each other, and in the relaxation of tension, doubled up with laughter. Mhoja described how he had dodged between the legs of the oncoming elephants and round bushes and had to run after Simeon, trying to stop him. Simeon's one thought was to get to the escarpment and climb up it and out of this horrible place with so many wild animals, and it took Mhoja a mile to catch him. They were starting on their way back to see what had become of us when they heard my shout.

In the excitement of this attack I had been so engrossed with the elephant activities that I had not even looked at their features in any great detail.

"Who were they?" Mhoja asked.

It was dreadful to admit it, but I did not know. I could only make a pretty good guess. Only once before had I seen an attack that started with no threat display in total silence, and that had come from four equally large cows. It must have been the Torone sisters. The silence of the attack was uncanny.

Green Penis Disease

JOYCE POOLE

As a child, Joyce Poole had gone to Amboseli on an elephant-viewing safari. She was an American who, having spent her early years in the United States, moved at the age of nine with her family to Nyasaland (now Malawi) while her father directed the Peace Corps program there. That time in Malawi, followed by four years living as a young adolescent in Kenya, made her feel that Africa was more her home than America; and when, in 1975, her father was hired to run the African Wildlife Foundation office in Nairobi, Poole took a leave after her first year at college in order to study elephants. She had acquired a general sense of them as intelligent and socially complex animals from reading Iain and Oria Douglas-Hamilton's book *Among the Elephants*, and during that first summer in Nairobi she met the Douglas-Hamiltons and an elephant researcher named Cynthia Moss. Moss invited nineteen-year-old Poole to volunteer at her project in Kenya's Amboseli National Park.

Moss had by then already identified a large number of the adult female elephants at Amboseli and was able to recognize many of them quickly. The adult males were more difficult to identify in part because elephant society is mostly based on female social associations—friendships, kinships, alliances—whereas the males are less social. They are more independent and more competitive, play rougher games, and show less interest in babies. By early adolescence, around the age of fourteen, they simply leave their birth groups entirely: at that point they choose to wander by themselves, to drift along at the edge of

142

other family groups, or to form loose associations with other males. At Amboseli, it was hard to know where to find any particular adult male or to recognize his identity from knowing what family he was physically close to at the moment. And because the males could not usually be identified as belonging to any particular family or bond group, the identification system that Moss had carefully constructed based on alphabetically coded names never worked well for the adult males. Instead of naming them, she numbered them—with each identifying number preceded by an "M." The males she became more familiar with were sometimes randomly given names that might have mnemonic value, but numbering was always the primary means of male identification.

By the time Poole joined the project, Moss had made tentative identifications of some sixty-eight adult males. Yet the identifications were often old, uncertain, and sometimes wrong. Moreover, most of the already identified males were rarely seen or had gone missing altogether. (It would later become clear that most of the older males had been killed by poachers looking for ivory.) At the same time, Moss was just more interested in the females and believed there was still a good deal to learn about female social and sexual behavior, so Poole agreed to study the males.

Aside from the genuine difficulties of finding and recognizing individual males, there was the additional problem that they seemed more dangerous than the females. Part of the added danger had to do with size. Elephant females stop growing during their midtwenties, while males continue to grow well into their forties, which means that adult males ordinarily get to be twice the size of adult females: an average of six tons rather than three, nearly thirteen feet tall at the shoulder instead of nine. Adult females are big, of course, and when threatened they may gather into a coordinated phalanx of flesh and forward-thrusting weaponry. But the males become mountainous, and they can also be, as Poole discovered, unpredictably aggressive.

As she was developing the identification catalogue of males, Poole also started to notice that some of them were coming down with what seemed like an infectious disease. The symptoms included swollen and secreting temporal glands and a persistent urinary incontinence: a dribbling so routine and marked that it turned the penile sheath "a greenish color from what seemed to be a nasty fungal growth." It was alarming, and since the green of the penis sheath was the most obvious marker, she tentatively named the affliction Green Penis Disease (GPD). She first noted the syndrome in February 1976, when an enormous male she called Zeus passed through their camp. Several weeks later, she observed a second male who was displaying similar symptoms. She named him Green Penis. The syndrome was simultaneously puzzling and troubling, and for a time she and Moss both worried that it could be the result of a venereal disease passing through the Amboseli population. In September of that year, Poole returned to America for her second year of college, receiving regular updates in the mail from Moss, who was seeing other big males—Dionysius, Aristotle, Agamemnon, David—with the same syndrome: secreting temporal glands and a persistent urinary incontinence resulting in a green penis. And when Poole returned to Africa for summer 1977, Moss added the intriguing observation that the green penis males also seemed to be sexually active and extraordinarily aggressive. She warned Poole particularly about a huge male with a deep V-shaped cut in his lower right ear. His steadily growing notoriety eventually earned him the name Bad Bull.

Another piece of the puzzle came into focus when Poole realized that some of the GPD males whom Moss had described a few months earlier as being sexually active and seeking fertile females were now not showing the slightest interest in females. Instead, they were relaxing and feeding in the company of other males. Meanwhile, Bad Bull and another GPD male named Cyclops had begun acting obsessed by sex and pursuing various females.

Poole began to wonder whether the putative disease had something to do with male sexual cycles. She mentioned the thought to Moss, who explained that all the experts agreed that while elephant females indeed pass through somewhat standard mammalian estrus cycles, males did not have sexual cycles. Still, both women had by then decided that what they originally thought of as a disease might be something closer to a normal cyclical syndrome of some sort. They stopped talking about Green Penis Disease and began speaking about Green Penis Syndrome. In January 1978, while she was in Nairobi, Poole attended a party where zoologist Harvey Croze showed her a scientific paper on musth in Asian elephant males caused by periodic surges in testosterone and resulting in hyperaggression, obsessive sexuality, a steady discharge from the temporal glands, and urinary incontinence. Poole recognized instantly, as she has written in *Coming of Age with Elephants,* that she and Moss had made the "very exciting discovery" of musth among African elephants.

O ne afternoon in February 1976 a group of females came through the North Clearing and into the camp, followed by an enormous male whom I had never seen before. His symmetrical tusks were thick and beautifully curved, and he carried his huge head carefully, as if its great weight might throw him off balance. He walked through camp with his head and ears high, towering over the other males, the females, and the calves. I had never seen such a large elephant before, nor do I think I have ever seen one as immense since. As he passed the dining room tent I noticed that urine dripped continuously from his penis. I took a closer look through my binoculars and realized that there was something terribly wrong with him. The constant dribbling of urine had apparently caused the sheath of his penis to turn a greenish color from what seemed to be a nasty fungal growth.

Studying the male again, I estimated that he must have been at least fifty years old. Perhaps old elephants become incontinent, I thought to myself at the time.

I saw the huge male again on the afternoon of February 22 feeding in the young regenerating acacia on the southern end of *Longinye* swamp. I now noticed that the sides of his face were marked by a dark stain of secretion oozing from swollen temporal glands. The temporal glands, located just behind the eyes, were still an enigma to African elephant biologists, and I wondered what was causing them to secrete so profusely. He was still dribbling urine, and his penis was still green.

Ten days later Cynthia and I were out watching elephants together when we saw him for a third time, again with a group of females. Looking back through her notes, Cynthia noted that she, too, had seen him before, once on the twelfth of February with two other males and again on the twenty-third with a group of females. On both occasions he had been dribbling urine. We saw him twice again in the following week. In honor of his magnificence, I named him Zeus and gave him an identification number, M103. But we never saw Zeus again: the last in a line of huge Amboseli males, he, too, probably succumbed to the ivory poachers.

Several weeks later I came upon another big, beautiful bull whom I had never encountered, standing with a group of females in a small patch of regenerating acacia near a favored bull area I had named Place of the Bulls. He, too, had a green penis and secreting temporal glands. He, too, was walking tall and sedately, closely following a group of females. He was in his late thirties or early forties and had long, curved, asymmetrical tusks. There was nothing remotely old or senile about this male. Cynthia and I saw him several times in and around *Oltukai Orok*, in the company of different groups of females, and once showing particular interest in an estrous female. I gave him

a number, M117, and named him Green Penis, for his terrible affliction. After a few more sightings, just as suddenly as he had appeared,
Green Penis disappeared again.

Long discussions developed in camp about the elephants' malady.
Other researchers and visitors suggested that the green penis was a
symptom of a disease, perhaps a form of elephant venereal disease.
This seemed plausible: certainly both Zeus and Green Penis had been
with females and appeared to be sexually active. For a time, the affliction became known as Green Penis Disease, or GPD, and although
we joked about it, I got the impression that the possibility that GPD
might be sexually transmitted made Cynthia slightly concerned about
her females and me, in turn, defensive about *my* males.

In September 1976 I returned to the United States to continue
my education at Smith College. Cynthia wrote to me with Amboseli
news, telling me that during the first half of 1977, she had recorded a
series of additional males with Green Penis Disease: first Dionysius,
then Aristotle, followed by Agamemnon, and David. Each of these
males had been in association with females.

When I returned to Amboseli for several weeks during summer
1977, Cynthia reported that some of the males with Green Penis
Disease were extremely aggressive and warned me to be particularly
wary of a male with a deep V cut in the lower part of his right ear. She
had been out watching a group of elephants south of Observation
Hill, when this huge male came and towered over her car in a very
threatening manner. According to her notes, this new male with the
V cut had temporal gland secretion. She moved off to another part
of the group but, to her dismay, he followed her and went for her
again, this time at full charge. It was unusual behavior for the usually
placid elephants of central Amboseli, and she was more than slightly
unnerved. As she drove away, she remembered back to her days in
Manyara and decided to follow Iain Douglas-Hamilton's advice and

to teach this bull a lesson: the next time he charged, she would charge him back. So, when for the third time he began to move toward her, she drove her Land Rover straight at him. To her horror, instead of turning and running away, he came for her at a full charge. They were on a collision course, and it was clear that the bull was not going to give way. At the last second Cynthia veered off across the open plains, her foot flooring the accelerator.

Deciding it best to avoid the nasty bull completely, Cynthia moved to watch another group of females who had gone off in the direction of the swamp. She drove across the plain and found a way through the line of young acacias down to the edge of the swamp. In those days the vegetation was a thick, tangled mass of thorns and branches, and there were only a few places wide enough to allow a vehicle through and onto the strip of grass along the swamp edge. Engrossed in her note-taking, she forgot about the big bull and sat quietly watching the elephants for close to an hour when she felt a shiver run down her spine.

Instinctively she looked in the rearview mirror; a pair of thick tusks and a huge gray mass filled the view. She knew immediately that this was the same animal and that he had gone out of his way to follow her down to the swamp. Without a moment's hesitation she started the engine, and, with the male charging after her, she found her way through the thick trees and out to the safety of the open plains on the other side. Later, when she had calmed down, she went back to look at the path she had taken: There she found his huge footprints in between the tracks of her tires and a squiggly line made by his trunk as it moved back and forth following her scent down to the swamp edge. The male with the V cut had clearly established his dominance, something he never again let any of us forget. Over the years his behavior became so legendary that he earned the name Bad Bull and was given the reference number M126.

During that same summer I began to notice that my favorite males were behaving in what seemed a very strange manner. Although Cynthia had written to me that Dionysius, Aristotle, Agamemnon, and David had been in the company of females, when I returned to Amboseli I discovered the first three completely uninterested in sex, choosing instead to spend their time with other males resting and feeding around *Oltukai Orok*. And I eventually found David, whom Cynthia said had completely disappeared, in *Olkelunyiet* on the eastern boundary of the park, also with a group of males, but ones I had never seen before. That summer Bad Bull and Cyclops instead were chasing after females and threatening other males.

Could it be, I began to wonder, that the Amboseli males had individual sexual cycles? In Cynthia's chapter on elephants in *Portraits in the Wild*, which summarized what was known about African elephants at the time, there was no mention of it. In those days I was rather intimidated by Cynthia, who knew so much more than I did about elephants and about life, and I pondered my theory for a long time before I ventured to suggest it to her. She listened, but told me that many people had studied elephants, and none of them had found any evidence to suggest that males had sexual cycles. I was disappointed, but I still believed that there was a distinct possibility that they did.

I returned to Smith College for the fall of my junior year, and Cynthia later wrote to tell me of two more males with the green penis affliction, first Hulk, then Green Penis himself. As with the others before them, each of these males was seen in the company of females and behaving very aggressively toward other males. By this time Cynthia and I had decided that GPD was not a disease but a behavioral syndrome, so we renamed it GPS.

I returned to Kenya for Christmas and spent January 1978 in Amboseli. There I found Dionysius, a male I had seen earlier only with other males, accompanying females and with Green Penis Syndrome. Just before I departed for Smith, I attended a lunch party at the National Museums of Kenya, where a friend of Cynthia's, Harvey Croze, handed me a paper that he thought might interest me. It was titled "Plasma Testosterone Levels in Relation to Musth and Sexual Activity in the Male Asiatic Elephant." I flipped through its pages until suddenly a photograph caught my eye: It pictured a male Asian elephant marked with two large arrows. The caption read: "Asiatic elephant bull in musth. The male is fastened to trees by chains around his front and hind legs, and is showing some discharge from the temporal glands behind the eye (arrowed) and a dribble of urine from the penis (arrowed)." I knew immediately that we had made a very exciting discovery in Amboseli. Our males with Green Penis Syndrome were neither incontinent, nor were they suffering from elephant VD, they were in musth—and we were the first to document it in African elephants.

I had heard about musth and knew that among Asian elephants it was a period of extreme aggression that occurred once a year in adult males, lasting a few weeks or months. It was associated with a dark secretion that oozed from the temporal glands; only male Asian elephants exhibited this secretion, and it occurred only during the period of musth. For this reason the temporal glands themselves were often referred to as musth glands. Musth was assumed to be a period of intense sexual and aggressive activity, or rut.

While musth was well documented in domesticated Asian elephants, strangely it was thought not to occur in African elephants. All the pioneering researchers of African elephant behavior and reproductive biology (including Roger Short, Erwin Buss, Silvia Sykes, Ian Parker, Richard Laws, Harvey Croze, Keith Eltringham, and Iain Douglas-Hamilton) had looked unsuccessfully for musth in African elephants and concluded that it did not occur. They based their conclusions on the fact that while temporal gland secretions were observed frequently in African elephants, unlike in their Asian cousins they occurred year round and in animals of all ages and of both sexes. The secretion appeared, therefore, to be totally unrelated to sexual and aggressive activity. What they had not realized was that African elephants exhibit more than one type of temporal gland secretion, and that the secretion by males associated with urine dribbling was, in fact, musth.

For some reason most African elephant biologists had overlooked the dribbling of urine. While Iain Douglas-Hamilton had mentioned it in *Among the Elephants*, he did not make the connection between urine dribbling and the period of musth. Looking back through her notes, Cynthia found a record from 1974 when she, too, had commented that Cyclops had been dribbling urine.

Our discovery that African elephants did come into musth was very exciting, and I made plans to spend the summer of 1978 in Amboseli collecting data on musth for my honors thesis at Smith. It was the first real contribution I had made to the body of scientific knowledge on elephants, and I was terribly proud and eager to continue.

Sex

Until the Amboseli study, very little was known about elephant sexual behavior. During his pioneering four and a half years at Manyara, Iain Douglas-Hamilton had witnessed mating only four times, and there were no other significant scientific accounts on the subject. A paucity or obscurity of elephant sex was accepted as the norm, which oddly enough may have seemed to confirm the classical idea that elephants were just very modest animals who preferred to do such things privately and in the dark. Cynthia Moss gained a similar impression about the obscurity of elephant sex during her first years at Amboseli, rarely witnessing anything that resembled mating behavior—except for the one day in 1974 when she observed a dozen males chasing a single female until a large male "caught" and mated with her four times. As it turned out, Moss's early years at Amboseli coincided with an extended drought, which ended in 1977. Once the rains returned, the vegetation proliferated, which brought a burst of nutrition followed by a burst of fertility. The world was once again ready for babies, and the females returned to their normal fertility cycles. It was a period when, as Moss has written, there was "such a profusion of sexual activity that at times it seemed like a bacchanalian debauch."

Suddenly, Cynthia Moss and Joyce Poole were surrounded by a wealth of new observations having to do with elephant sex and sexuality. In the years 1977 and 1978 alone, Moss saw nineteen matings. Moss, meanwhile, looked for "odd nuances of behavior" associated with sexual activity among the adult females at Amboseli. Eventually she acquired a set of more than a hundred observations that seemed to indicate visible signs of estrus, including a certain "wariness" and posture, and a style of movement she called the *estrous walk*.

Meanwhile, the discovery of musth among African elephants at Amboseli was accompanied by an expanded understanding of male sexuality and the dynamics of male competition over access to fertile females. In some ways, the primary social unit of elephants—the matriarchal family unit—may seem to humans like an ideal society. The matriarch maintains order mainly through her status as the wise elder of the group rather than through force, and the group as a whole persists through the power of positive attachments. Mothers are not alone in caring for their offspring, since aunts and aunties also keep watch over the young while older sisters will watch over and play with their younger siblings.

Once the males leave their birth family at adolescence, they enter an all-male society that may seem a good deal less welcoming and less coherent. Males sometimes move near or with other males, and sometimes they wander alone. Their relationships with one another, although often tolerant and even friendly, are also competitive and ultimately serve to establish a dominance hierarchy. Dominance among males is generally based on size, and size is a good indicator of age, since males steadily grow bigger for most of their lives. Males between the ages of twenty and thirty will have reached the size of adult females, and as they become larger they almost automatically acquire greater dominance. Males over thirty will have achieved an undeniable status among the others, and a female will not challenge that status; at the approach of such an older male, she is likely to present a friendly or submissive greeting. The transition into a musth state also is affected by age, since the younger males experience it less intensively and for shorter times, while the older males can enter musth states that last for weeks or even months. But musth is the wild card, since the sudden massive surge in testosterone levels changes a male's character in a fashion so swift and intense that it has been described as a "Jekyll and Hyde" transformation. Suddenly a gentle male who spent his days in casual contact with others resting, relaxing, feeding, and wallowing leisurely in the mud is transformed into a giant beast obsessed by sex and overcome by a profound antagonism toward all other males.

Fertile females advertise their presence through the condition of estrus, and a musth male spends his days wandering from one family to the next, guided largely by his ears and nose; and his nose is possibly the most exquisitely sensitive olfactory organ on this planet. He is searching exclusively for females who have ovulated and are therefore fertile. Once he reaches an elephant family group, he may quickly discover which females are fertile or, less quickly, he may move from one to the next, testing their vulvae and urine in a distinctive manner that involves a dedicated olfactory system—known as the vomeronasal organ—that is responsive to the chemistry of fertility and connects directly to a primitive part of the brain associated with sex and aggression. Testing involves taking a fluid sample with the tip of his trunk, raising the sample to his mouth, and grimacing—the so-called Flehmen response—in a fashion that opens the vomeronasal passages.

Saibulu and Chris, medium-sized bulls in their mid-twenties, raced at full speed in pursuit of Tia. She wove in and around tree stumps and bushes and finally reached open ground, where she took off running as fast as she could. The two males followed close behind, but they did not seem to be as fast as she was or perhaps as determined, and she began to draw away from them. Tia made a large arc and by the time she reached its apex she was thirty yards ahead of the males. She slowed down but continued walking fast and at the completion of the arc she arrived at the point where Slit Ear, Teresia, and the others had reached. The males, standing out on the plain, had given up and were temporarily resting from their sprint.

When Tia returned to her family, the others rumbled at her arrival. Her two-year-calf rushed up to her and opened his mouth and let out a hoarse bellow. She touched his mouth with her trunk and he moved closer and began to suckle. Tallulah seemed particularly excited and came over and reached her trunk toward Tia and smelled her and rumbled and then shook her head.

Ten minutes later another medium-sized bull, Mac, approached the family and began to test the females. When he attempted to test Tia she moved off at a fast walk, with her head held up, her ears lifted and slightly spread, looking back over her shoulder at him. He immediately abandoned the other females he had not yet tested and started after Tia. He released his penis from its sheath and began to get an erection. Tia walked out from the family and Mac followed at the same speed, keeping about 20 yards behind her. When they were about 150 yards away from the others, Tia stopped, and, still at a distance of 20 yards, Mac stopped too. Tia stood for a while picking the dusty soil up in her trunk and spraying it onto her head and back. She was standing with her back toward Mac but at a slight angle so that she was able to watch him. He took a few steps toward her and immediately she began to walk away. They continued on around in a circle until she came back to her group.

By now the other two males, Saibulu and Chris, had returned to the TCs and TDs and two younger males had joined them. Tallulah came out to meet these young males and briefly engaged one in a sparring match. Afterward she rubbed her head against Chris's hindquarters and then stood among the five males. She seemed agitated and alert. The males, however, were not interested in her. When Tia had been back in her group for only a few minutes she was chased by all three of the medium males. She ran flat out again and managed to get away.

The sequence of either running or walking after Tia occurred many more times in the course of the morning, but none of these males ever got close enough to touch her. Sometimes her calf ran with her when she was chased, desperately trying to keep up. Eventually he learned to stay back with the other females. When the group stopped in the shade of an acacia tree, the bulls rested as well. Later in the afternoon the pursuit resumed, and at 4:30 Tia was finally caught by Chris in thick habitat where it was difficult to run. Chris managed to stop her and mount, but Tia kept moving sideways and he was unable to penetrate. At the same time Tia's family came running over and, surrounding the pair, they rumbled and trumpeted. Chris ejaculated all over the back of Tia's legs and dismounted.

Tia stood with her head low, looking tired after all the running she had done during the day. She had had very little time to feed and was harassed even when she tried to drink. Soon most of her family drifted away to begin feeding again, leaving Tia and Tallulah resting. Chris also stood quietly nearby. Tallulah remained alert and was the first of the females to hear the low rumbling sound almost like an engine far in the distance. When Tallulah rumbled in response, Tia listened and could also hear the distant sound. They waited tensely, listening, and after about ten minutes they could smell the sharp, acrid, unmistakable odor of a musth bull—a large, sexually active male elephant in full rut.

Tia and Tallulah watched the arrival of Bad Bull with great interest. He was a massive animal with two jaggedly broken tusks and a large V notch out of the bottom of his right ear. About forty-five years old, he stood at least two feet taller than the medium-sized bulls and his head, particularly his forehead and the space between his tusks, was extraordinarily broad. His temporal glands, one on each side of his face, located midway between the eye and ear, were grotesquely swollen and secreting a copious, viscous fluid. His smell, almost intoxicating to the females, originated from the temporal glands and from the area of his penis and sheath. He was dribbling urine continuously, spurting it out at a very fast rate. His sheath was covered in a greenish white scum and his inner legs were black with wetness.

Even though they had all heard and smelled him coming, his arrival caused a considerable commotion. He walked straight for the females, carrying his head high, tucking his chin in, and waving his ears in a characteristic musth male posture. He did not bother to drape his trunk over his tusk and lower his head to reassure the females. They were both attracted and alarmed. First they rumbled a greeting to him, and as he got closer they milled about, turning and backing, some reaching their trunks toward him, others urinating. Even the young calves were fascinated and came over to smell him. The males who had been surrounding Tia, more or less, waiting for her next move, immediately ambled off to the periphery of the group, where they started to feed, "pretending" that they had no interest in Tia, but all the time positioning themselves so that they could keep an eye on her and Bad Bull.

Bad Bull did not test the other females, even though Tallulah kept placing herself right in front of him. Tia had walked off a few paces in the characteristic estrous posture of ears held out tensely, head up and turned slightly, looking back over her shoulder, and he noticed her immediately. He turned toward her and approached. As soon as she had his attention, she began to walk faster, heading away from the others. He too picked up speed and at the same time his penis emerged from its sheath and dropped down, nearly touching the ground.

Tia began to run, but not fast and not with the determined intensity with which she ran when chased by the younger bulls. Bad Bull quickly caught up with her. He reached out with his trunk and placed it along her back. She stopped immediately. He moved his trunk forward to the top of her head until his head and tusks were over her hindquarters. Then, using his head as a lever, he reared up, placing his front feet on her backside but sinking back so that nearly his whole weight was on his hind legs. His penis had hardened, changing in shape from a simple arc to an S-shaped double bend. It curved out from the sheath, then down, and then curved back up at the tip. Fully extended and erect, it measured almost four feet and was endowed with musculature allowing directional control. Bad Bull whipped it about, searching for Tia's vulva, which hung low between her legs with the opening facing the ground, not up under her tail as in most ungulates. Tia backed in closer and spread her legs. The S-shaped curve of the penis now came into play. Bad Bull sunk down even farther and hooked the tip of his penis into the opening and thrust upward, forcing the full length of the penis into the vagina, at the same time lifting the whole vulva and protuberant clitoris at least a foot. Tia stayed still and they held this posture for about forty-five seconds, after which Bad Bull dismounted and withdrew his penis with a gush of semen.

Tia stepped forward with legs still spread, opened her mouth, and uttered a very deep, loud sound, more like a bellow than a rumble. Her family came running over, flapping their ears and rumbling, screaming and trumpeting in a great outburst of vocalizations. They extended their trunks toward her, some to her mouth, others to her genitals. Tia turned back to Bad Bull and reached her trunk in the direction of his penis, which, although no longer fully erect, still hung down free of the sheath. She then turned and stood parallel to him and rubbed her head against his shoulder. The members of her family began to calm down but Tia continued to give repeated low, long, pulsated rumbles for the next ten minutes, and occasionally reached her trunk to touch Bad Bull's penis or to smell the sperm on the ground. Tallulah remained close to the pair and still showed some excitement. After a while, she moved in close to Bad Bull and stood on his other side.

Bad Bull stayed with Tia for the rest of the afternoon and all through the night. At the same time Tia made sure she was never more than a few feet from him. With him there she could feed and drink and test without being harassed by the other bulls. They did not mate again until the next morning and this time Tia did not even run from Bad Bull to start the sequence, but once again her family raced over with excitement and Tia made the long, low postcopulatory rumbles afterward. By morning there were eight other bulls with the TCs and TDs, all being very discreet about their interest in Tia, in fear of reprisals from Bad Bull. Several times in the night a male would try to inch in close to the pair, and Bad Bull would lunge at him, sending him running off with a groaning sound. A few times Tia wandered a couple of yards away while feeding, but as soon as a male tried to approach she moved straight back to Bad Bull.

They stayed in consort throughout the next two days and mated only one more time. By the fourth day Bad Bull began to lose interest in Tia. She tried to maintain the consort relationship with him but he did not guard her with the same enthusiasm and eventually he allowed the other bulls to separate her from him. Soon after, he left the TCs and TDs and strode off in search of other estrous females. Without a protector, Tia was once more harassed and chased by the younger bulls. But once again Tia was determined to get away from them and although they were relentless in their pursuit, she was able to evade all but one of them. Hector, a fairly big bull, managed to run her down and mate with her, but the chances of being impregnated by him were low, as she was now at the end of her estrous period. By the following morning the bulls had lost interest in her and they too left the family. Tia returned to her confused and weary calf and resumed the normal routine of elephant family life.

EMOTIONAL
AND COGNITIVE
ELEPHANTS

Joy

CYNTHIA MOSS

Cynthia Moss, an American journalist working for *Newsweek*, came to Africa as a tourist in 1967 and happened to visit Lake Manyara National Park. Inspired by meeting Iain Douglas-Hamilton, she quit her job in America and moved to Africa to volunteer as an assistant on the elephant project. Moss spent eight months working with Douglas-Hamilton in 1968 and returned for briefer periods until 1972, when she began her own elephant research project at Amboseli National Park in southern Kenya.

Like Manyara, Amboseli is among the best places in the world for viewing wild elephants. A 150-square-mile pastiche of dry savanna, woodland, papyrus swamp, and salt pan dominated by the halluci-natory vision of a looming, snow-topped triangle known as Mount Kilimanjaro, Amboseli contains antelopes of many species as well as wildebeests, zebras, buffaloes, giraffes, hyenas, lions, cheetahs, leop-ards, rhinos, and several hundred elephants. The Maasai people have lived in this part of Kenya for the past four hundred years, so Amboseli is marked by Maasai place-names (the Ol Tukai Orok woodlands and the Enkongo Narok and Longinye swamps, for example) and protected by a traditional Maasai attachment to cattle and disinterest in hunting, except for those occasions when a young warrior may seek to demon-strate his courage by spearing some dangerous animal, such as a lion, buffalo, or elephant.

When she began her study at Amboseli, Moss was already familiar with Douglas-Hamilton's methods for identifying individual elephants: taking photographs and assembling identification catalogues
that concentrated on the ears, which are an elephant's most obvious
equivalent of fingerprints for humans. She also relied on other markers
of individuality, such as the size and shape of tusks. Eventually she
came to know the elephants at Amboseli well enough that even from
a significant distance she could look at an individual and understand
from such characteristics as body shape and style of movement who it
was—a recognition much like, as she has written, "recognizing a human
friend who is walking away from you on the other side of the street."

As Douglas-Hamilton had found with the Manyara elephants, Moss
noted that the Amboseli elephants lived in matriarchal societies. The
primary social groups were based on critical bonds between closely
related adult females with one another and with their offspring of
both sexes. Since males invariably left or were driven out of their birth
groups after the start of adolescence, these family units or groups included no adult males and were led by the oldest of the adult females.

Douglas-Hamilton had observed that the elephant family groups
at Manyara maintain consistent preferential associations with certain
other family groups, and he called the larger multifamily associations *kinship groups*. Moss found a similar social structure among
the Amboseli elephants, but she preferred to call them *bond groups*.
Individual members of the bond groups often demonstrated their
strong ties with one another through many friendly and supportive
behaviors: for example, mothers comforting each other's infants, individuals casually leaning against one another, and everyone grouping
together in response to threats. Most remarkably, members of the
same bond group would celebrate the reuniting of their families after
a time apart with what Moss called a *greeting ceremony*: a spectacular
demonstration of clicking tusks, trumpeting, screaming, and spinning

around while urinating, defecating, and secreting copious fluids from their temporal glands (behind the eyes).

Social relationships among elephants are intense and complex, in short, and Moss designed a system of alphabetical identification to help simplify the complexity for human observers. There were more family groups at Amboseli than letters of the alphabet, and partly for that reason she assigned each family a pair of letters—starting with AA, AB, and so on—and then, usually, she identified each individual in the group with a name that began with the family's first letter. Thus, for example, the TD family, led by the matriarch Teresia, included her three older daughters Trista, Tina, and Theodora, her young son Tolstoy, and her younger grandchildren Tallulah and Tim. Families that were part of the same bond group would be designated with a common initial letter—so the TD family led by matriarch Teresia was associated in the T bond group with the TC family that included two older females Tia and Tess and the matriarch, Slit Ear. Yes, there were exceptions to the alphabetical system, usually inspired by some individual's unusually distinctive marker.

More than any previous commentator, Moss came to recognize and describe the powerful *emotionality* of elephants. Members of a family or a bond group, she saw, communicated with one another with postures and gestures and vocalizations that clearly expressed strong feelings of affection and attachment, and she understood that their social systems cohered over time largely because of the ties created by emotions and emotional signaling. The following excerpt from Moss's book *Elephant Memories: Thirteen Years in the Life of an Elephant Family* is remarkable for its clear description of elephants as emotional animals—and also for its presentation of elephant behavior in the absence of a human perspective. The author has artfully removed herself from the story and worked to bring readers as close as possible to the reality of elephants in an elephant world with elephant values, including, in this case, the powerful emotional attachments between the TD and TC families led by the matriarchs Teresia and Slit Ear.

——— —— —— ——   —— —— —— ——

Teresia and her two calves, Theodora and Tolstoy, and her grand-children, Tallulah and Tim, walked south, first through a strip of *Acacia tortilis* trees and on through an area of lava rocks and small bushes. It was early morning, an unusual time for them to be moving in that direction. In the late afternoon of the previous day they had left the Longinye swamp with Slit Ear and her family, had crossed into the Ol Tukai Orok woodlands, and fed on the palms and *Acacia xanthophloea* trees for a while before settling down to sleep in an open area near the southern tip of the Enkongo Narok swamp.

Slit Ear had woken earlier than Teresia and had turned back toward Ol Tukai Orok and Longinye in the traditional pattern of moving out from the swamps at night and back to them in the daytime. She and her family were nearly a mile away when Teresia woke and stood up. The rest of her little group got up as well, stretched and yawned, and dusted and scratched themselves. Teresia gave a long, low rumble, the contact call that basically says, "Here I am, where are you?" She held still, listening, her head high and ears extended, but heard no answer. She called again, this time making an even lower-frequency sound that traveled farther. She heard a distant answer from Slit Ear, and turning her head to locate where Slit Ear was, Teresia rumbled again and once more got an answer.

However, instead of heading off in the direction Slit Ear had taken, Teresia turned toward the south. The rest of her family remained facing north. Teresia swung her right front foot back and forth in the characteristic elephant gesture of indecision. She took a few steps forward, picked up some dust, and threw it on her back, called again, waited and listened. Then she decisively lifted and spread her ears, flapped them against her neck and shoulders, let them slide down with a rasping sound, and strode off toward Kilimanjaro. Her family loyally followed.

The TD family walked south for about a mile and a half without stopping to feed. Teresia seemed to have a particular destination in mind. Just before dawn they arrived at a wide depression more or less at the base of the mountain. This area consisted of small, thorny trees, *Balanites glabra* and *Acacia nubica*, interspersed with some grasses and low bushes. It was a favored feeding area for the Amboseli elephants in July and especially August. These were mid-dry-season months when the more palatable and nutritious grasses in the bushed grasslands were finished but there was still good forage left in some of the woodlands outside the park. The elephants used the woodland to the south before finally retreating once again to the swamps within the park for the late dry-season months of September and October. While feeding there they had to walk back to either Longinye or Enkongo Narok swamps to drink, but they usually did this at night. Teresia, who was in her late fifties, had been using this area in July and August from the time she was a young calf. Following the tradition of her family she had decided to spend a few days in the south.

For three days the TD family fed in the depression, drinking at night or in the early morning, but they were never completely relaxed there. There were many Maasai around, and the elephants had to be alert to their movements and activities. Still the food was good and nothing unpleasant happened. But the loss of Trista had reduced the size of the TD family, and they did not feel as secure as they would have in a larger group with just that many more eyes and ears and trunks on the alert. Without the eight members of Slit Ear's family for company and added protection, they were isolated and vulnerable. On the morning of the fourth day Teresia led her family north, toward Amboseli and Slit Ear.

When they crossed the park boundary and found themselves back in the Ol Tukai Orok woodlands, the younger members of the family started running and trumpeting. Tallulah, Theodora, Tim, and Tolstoy let their bodies go loose as they ran, which made their ears and trunks flop about. At the same time, they trumpeted with a loud pulsating sound associated with play. Teresia more or less maintained her dignity but she too broke into a run. When the younger animals got ahead, Tim and Tolstoy stopped, whirled around, and came together for a sparring match. Tallulah and Theodora found some bushes and began bashing through them and beating them with their tusks. Their tails were curled up over their backs, their heads were held high with their noses poked out, and the whites of their eyes showed as they rolled the pupils up. Teresia caught up with the others and stood with her head up and out. In her usually placid brown eye was a decidedly wild glint.

The young animals played for another ten minutes before everyone began to feed on the regenerating acacias, breaking off small branches and carefully stripping them of the tasty and nutritious bark. After a while they switched to the grass that was growing under and around the trees. It had been a year of unusually high rainfall, and even though it was getting toward late dry season there was still some grass remaining. They fed for another twenty minutes and then all stopped at once when they heard an elephant give a contact call about a mile away. They listened for a moment but resumed feeding and did not answer because it was not someone from their bond group.

Soon after, they began walking slowly, feeding as they went, through the mixed acacia and palm trees, finally arriving at the northern edge of the woodland at midmorning. Here they rested, standing in a tight bunch under one of the yellow-barked trees. On reaching this spot Teresia called several times but got no response. She and her group were eager to rejoin Slit Ear, but they did not know where she was.

After their rest they crossed the pan and open plain and reached the northern end of the Longinye swamp. They fed there for the rest of the day. Teresia continued to call at intervals, but still got no response. All the members of the group had fresh temporal-gland secretions almost continuously. They met several other family groups in Longinye but ignored them, passing them by or feeding near them without interacting.

That night they moved out of Longinye to the east into the Olodo Are woodlands. They spent the night feeding and resting. Teresia called several times during the night and in the early hours of the morning when the group woke up. On that occasion she heard an answer more than a mile away toward the northwest but she was not sure it was Slit Ear. She started to move in that direction in any case. They walked west until they arrived at an area of regenerating acacia trees on the eastern side of Longinye. Here Teresia stopped and rumbled. She instantly got a response from about a quarter of a mile away and to the south. This was definitely Slit Ear, and Tia and Tess answered as well.

Teresia changed her course and began walking rapidly to the south. She and the others were streaming from their temporal glands and the sides of their faces showed thick black lines all the way to their chins. They walked through the patch of young acacias and emerged in an area of swamp interspersed with wooded islands. At a little rise above the swamp, they stopped and Teresia, Theodora, and Tallulah rumbled, listened, heard a response, and rumbled a louder sound. They were clearly excited, with heads held high and ears lifted. They shifted their course slightly and, almost running, plunged down into the swamp course and up the other side onto the first island.

Suddenly ahead of them was a group of elephants running out of the trees and coming straight for them. Teresia stopped for a moment in alarm, then recognized Slit Ear, and both groups ran, rumbling, screaming, and trumpeting toward each other. The younger animals had moved ahead of Teresia, but when the two groups came together, Slit Ear ignored the others and pushed through them to reach Teresia and greet her. Both elephants raised their heads up into the air and clicked their tusks together, wound their trunks around each other's while rumbling loudly, and holding and flapping their ears in the greeting posture. They whirled around and leaned and rubbed on one another. Meanwhile all the other members were greeting each other with much spinning, backing, urinating, earflapping, entwining of trunks, and clicking of tusks. All the elephants were producing so much temporal-gland secretion that it dribbled down along their chins and into their mouths. Above all, the sounds of their greeting rent the air as over and over again they gave forth rich rumbles and piercing trumpets of joy.

After eighteen years of watching elephants I still feel a tremendous thrill at witnessing a greeting ceremony. Somehow it epitomizes what makes elephants so special and interesting. I have no doubt even in my most scientifically rigorous moments that the elephants are experiencing joy when they find each other again. It may not be similar to human joy or even comparable, but it is elephantine joy and it plays a very important part in their whole social system.

Triumph and Grief

JOYCE POOLE

During the second half of the twentieth century, the new science of animal behavior took scientists out of the laboratory and into the field and brought them face to face with large and large-brained animals and the remarkable complexities of their behavior. Scientists began to recognize many kinds of species as made up of complex individuals who appeared to experience the world complexly: who made decisions, remembered the past, anticipated the future, and interacted with the world and with each other in ways that might best be explained by reference to humanlike emotional and cognitive systems.

The following excerpt from Joyce Poole's memoir *Coming of Age with Elephants* describes one of the first close observations of a wild elephant giving birth: with the new mother, Tallulah (granddaughter of the matriarch Teresia), soon joined by several other females from her bond group in a wildly excited gathering that seems like nothing less than a joyful celebration of the event. Poole then describes witnessing (at a later time and in the company of another researcher named Cynthia Jensen, or "Cyn") a contrasting situation. Young Tonie from the Tuskless family, having just given birth, is now standing a solitary watch over the body of her stillborn infant—and seems to be expressing a response that could be described as grief. "As I watched Tonie's vigil over her dead newborn, I got my first very strong feeling that elephants grieve," Poole declares. "I will never forget the expression on her face, her eyes, her mouth, the way she carried her ears, her head, and her body. Every part of her spelled grief."

Early one morning I was driving along the southern edge of *Oltukai Orok* when I came upon a group of eighteen elephants. It was Teresia's and Slitear's families, part of the T bond group. I stopped to take some notes and noticed that one of Teresia's granddaughters, a young female named Tallulah, was acting strangely, lowering her hind legs and dropping down on her knees. As I studied her behavior closely, I noticed that a slight bulge had appeared below her tail, that her vulva was swollen, and that she was urinating in a continuous dribble. The bulge under her tail confused me until I realized that it was moving slowly downward. The dropping down on her knees had been a contraction; Tallulah was about to produce a baby!

As I fumbled with my camera, my hands trembling with excitement, Tallulah reached the edge of *Oltukai Orok* and disappeared from view in a clump of Phoenix palms and regenerating acacias. I found her lying on her side under a palm at 8:38 A.M. Leaving my car, I crept slowly toward her so that I could get a better view. A minute later Tallulah stood up, and I noticed that the bulge had become larger and had descended farther down her birth canal. At 8:41, Tallulah looked directly at me and walked away into a patch of acacias, only to turn around and come back toward me. Two minutes later she walked away again and disappeared into the acacias. I drove quickly to the other side of the clump of trees to find her baby on the ground at 8:44, still contained inside its amniotic sac.

For a minute Tallulah stood calmly and quietly over her newborn, and then gently touched it with her forefoot. The baby kicked its tiny feet in response. Tallulah bent over and, using her tusks, freed the newborn from the sac. Then she began to scrape the earth with her front feet, clearing the vegetation away. Her baby moved again, and Tallulah became more excited, scraping at the ground with renewed vigor. At 8:47 some members of her family arrived, rushing in to surround Tallulah and her newborn, with their heads and ears high, urinating and rumbling loudly. Bending down and using her feet and her trunk, Tallulah tried to get her baby to stand. The other females gathered around, continuing to scream, rumble, and trumpet, as temporal gland secretions streamed down the sides of their faces. Tallulah's high-ranking aunt, Slitear, now made her entrance, backing into the group amid more urinating, rumbling, and trumpeting. Tallulah continued to scrape the ground, tried to lift her baby, scraped the ground again, and then gently touched the infant with her trunk.

From my position, fifteen meters away, the newborn was barely visible through the scores of legs, trunks, and tails. At 8:56 the baby was pushed to its feet but toppled over. Two minutes later it was once more on its feet, only to fall over again. It took the baby more than half an hour of struggling before it finally managed to steady itself on its four legs. New females continued to arrive, some from Tallulah's bond group and others from outside families. The excitement in the group was intense, and several females dropped to their knees and, with their heads held high in the air, flopped their trunks about and gave me a wild look, the whites of their eyes showing. It was as if, in their excitement, they had forgotten that I was not an elephant and wanted to share their emotion with me. Another young female picked up the amniotic sac with her trunk, waved it in the air, and then tossed it over her head.

Finally, at 9:21, forty minutes after its birth, the baby took its first steps. The group began to make its way slowly along the edge of the palms, leaving the birth site, an area eight meters in diameter, trampled beyond recognition. At 9:47 the newborn passed its meconium, or fetal stool, and at 10:00 it suckled from its mother. The new baby was a female and Cynthia later named her Tao.

Several months later, early one morning in the late dry season, Cyn had driven from camp to *Ol Tukai* and noticed an elephant moving in a strange manner, walking and dropping down on its knees, out on the plains beyond the palms. When I returned from the field later that morning I, too, noticed the single elephant, and I thought it strange for it to be standing alone on the plains in the heat of the day. When Cyn returned to camp just before noon, she said that she thought the elephant we had seen on the plains had a dead baby at her feet.

Cyn and I returned to the site together, and as we approached the elephant we realized that it was Tonie from Tuskless's family. She was still having contractions, and blood was dripping from her vulva, but her behavior was subdued and so unlike Tallulah's. She stood quietly, her head and ears hanging forward, playing slowly, gently with the afterbirth with her trunk. The newborn at her feet was dry, and she repeatedly nudged it gently with her feet and finally rolled it over several times. Two vultures waited nearby. Tonie stayed out on the barren plains with her dead baby for the rest of the day and through the long night.

The following morning Cyn and I left the camp on foot and walked to the edge of the palms from where we could see Tonie still watching over her stillborn infant. Fifteen vultures and a jackal hovered around her; she charged and they scattered for a few seconds, only to return. Tonie placed herself between her baby and the scavengers, and, fac-ing them, she gently nudged the body with her hind leg. As I watched Tonie's vigil over her dead newborn, I got my first very strong feeling that elephants grieve. I will never forget the expression on her face, her eyes, her mouth, the way she carried her ears, her head, and her body. Every part of her spelled grief.

By now Tonie had been standing out on the bare plains without food or water for over twenty-four hours. Cyn and I walked back to camp, found a jerry can, and filled it with water. Interacting with one's study animals, let alone providing water to elephants in national parks, was not really proper behavior for a scientist, but under the circumstances I didn't care.

As we drove toward Tonie she charged, and I stopped the car. I placed a basin on the ground, poured the water into it, and then drove away. She lifted her trunk toward the water and walked immediately toward it, pausing only once. She drank quickly, emptying the basin in two trunkfuls. I returned to fill the basin again as she stood close by.

Later that morning Cyn and I returned to Tonie with another two containers of water. As she saw me put the basin on the ground she walked over and stood by the car. I held the twenty-liter can on my lap and, with one leg on the ground, I poured water into the basin. Tonie drank while I poured the water onto her trunk, her tusks no more than ten centimeters from my head. After she had emptied both cans, she reached through the door of my car and twice touched my arm with her trunk.

In the early afternoon I returned once again with more water. She emptied the first container and then waited patiently while I banged around trying to get the second can from the back of the car. She drank most of this second jerry can, then used the last bit of water to splash herself. In all Tonie drank ninety liters of water. After she had finished splashing, she again reached inside the car and touched me gently on my chest and arm.

The following morning we found Tonie still on her vigil, attempting to chase away the ever-closer vultures. Later that day she had gone, and all that remained on the plains was a few vultures and scattered bones.

Big Love

CARL SAFINA

Cynthia Moss and Joyce Poole have been, I believe, correct in attrib-uting emotions to elephants. To some degree their view of elephant emotionality follows ordinary common sense. Almost everyone has had a passing experience—perhaps one of looking into the eyes of a chimpanzee in a zoo or of settling down beside the family dog in front of a fire in the fireplace on a cold day—where it seemed intuitively clear that the animal had an intelligence and emotions of some sort, perhaps even a coherent consciousness not radically different from human consciousness. Of course, those ideas could be wrong, and certainly, when considered as mere intuitions or the product of a simple, anecdotally recalled event, they remain in a different realm from scientific understanding. Only a generation or two ago, in fact, no reputable scientist would have professionally entertained the notion that animals can think and feel. Such was the ruling scientific paradigm. Humans are emotional and cognitive beings. Animals are not.

This conceptual divergence between human and nonhuman has deep roots reaching back at least to the time of Enlightenment philosopher René Descartes, who famously asserted that humans are separated from all other living creatures by an uncrossable chasm. As the philosopher wrote in his *Discourse on Method* (1637), animals possess "no reason at all," and therefore nature "acts in them according to the disposition of their organs, just as a clock, which is only composed of wheels and weights, is able to tell the hours and measure the time."

Unreasoning and unconscious, then, animals are merely naturally oc-
curring machines.

With *On the Origin of Species* (1859), Charles Darwin challenged
Descartes's radical discontinuity directly and emphatically. Darwin
argued that the unmistakable tree of anatomical similarities already
identified among species was actually a family tree, a representation
of genetic relationships; and with his theory of evolution by natural
selection, Darwin convincingly explained how such a family tree could
develop over extended time. The concept of evolution by natural selection would be accepted by most biologically oriented scientists
during Darwin's lifetime as the best explanation for anatomical continuity among species. It would take more than a century, however,
before Darwin's understanding of the logical sequel to anatomical
continuity—that is, neurological, behavioral, psychological, and mental
continuity—would be considered with a comparable interest. At least
three important events have happened since the middle of the twentieth century to expand scientists' receptiveness to such ideas.

The first was an explosion in the numbers and quality of large
mammal observations in the field. Then came the publication of two
influential books by Donald Griffin, *The Question of Animal Awareness*
(1976) and *Animal Thinking* (1985). Griffin had earned his reputation as
a brilliant experimental zoologist by discovering and documenting that
bats navigate by echolocation. His two early books combined rigorous
and compelling argument with an almost encyclopedic array of references to animal behavior that showed evidence for thinking, planning,
awareness, and creative problem-solving. The third important event
is emerging in contemporary time. That is the application of modern
technologies and techniques of neuroscience to the comparative study
of human and nonhuman animal brains. These comparative studies
have begun reinforcing the concept Darwin always maintained: that
evolutionary continuity among related species will inevitably extend

to the emotional and cognitive. Animals may not have human-style language, but in a nonlinguistic fashion they think, feel, and are aware.

Scientist and author Carl Safina has written about this paradigm shift in *Beyond Words: How Animals Think and Feel* (2015), the first third of which focuses on elephants. Safina spent time at Amboseli, and his own descriptions of elephants as thinking and feeling creatures are based largely on that visit and some extended interviews of Cynthia Moss and others, including Amboseli researchers Katito Sayialel and Vicki Fishlock. The following excerpt is taken from a brief encapsulation of the book that appeared in *Orion* magazine.

Alison, fifty-one years old, is right—*there*, in that clump of palms—see? And there is Agatha, forty-four years old. And this one coming closer now is Amelia, also forty-four. Amelia continues approaching until, rather alarmingly, she is looming so hugely in front of our vehicle that I reflexively lean inward. Cynthia Moss leans out and talks to her in soothing tones.

Cynthia arrived in Kenya forty years ago, determined to learn the lives of elephants. The first elephant family she saw, she named the "AA" family, and she named one of those elephants Alison. And there she is. Right there, vacuuming up fallen palm fruits. Astonishing.

With much luck and decent rainfall, Alison might survive another decade.

Amelia, practically alongside now, simply towers as she grinds palm fronds, rumbles softly, and blinks.

In the light of this egg-yolk dawn, the landscape seems an eternal ocean of grass rolling toward the base of Africa's greatest mountain, whose blue head is crowned by snow and wreathed in clouds. Through gravity-fed springs, Kilimanjaro acts like a giant water cooler, creating two miles-long marshes that make this place magnetic for wildlife and for pastoralist herders. Amboseli National Park got its name from a Maa word that refers to the ancient shallow lakebed—half the park—that seasonally glitters with the sparkle of wetness. The marshes expand and contract depending on the rains. But if the rains fail, panes of water dry to pans of dust. And then all bets are off. Just four years ago, a drought of extremes shook this place to its core.

Through times lush and calamitous, through these decades, Cynthia and these three elephants have maintained their presence, urging themselves across this landscape. Cynthia helped pioneer the deceptively complex task of simply seeing elephants doing elephant things. Longer than any other human being ever has, Cynthia has watched some of the same individual elephants living their lives.

Several happy elephants are sloshing through an emerald spring under ample palm shade. It's a little patch of paradise. With bouncy, rubbery little trunks, the babies seem to transit the outer orbits of innocence.

"Look how fat *that* baby is," I say. The fifteen-month-old looks like a ball of butter. Four adults and three little babies are wallowing in one muddy pool, spraying water over their backs with their trunks, then sprawling on the bank. As a little one melts in pleasure, I notice the muscles around the trunk relaxing, eyes half-closing. An adolescent named Alfre lies down to rest. But three youngsters pile on, stepping on Alfre's ear. *Oomph.* The fun softens to a snooze, with babies lying asleep on their sides, adults standing protectively over them, the adults' bodies touching one another's as they doze. Feel how calm they are, knowing their family is safe here now. It's soothing just to watch.

Many people fantasize that if they won the lottery, they would quit their job and immerse themselves in leisure, play, family, parenthood, occasional thrilling sex; they'd eat when they were hungry and sleep whenever they felt sleepy. Many people, if they won the lottery and got rich quick, would want to live like elephants.

The elephants certainly seem happy. But when an elephant seems happy to us, do they really feel happy? My inner scientist wants proof.

"Elephants experience joy," Cynthia says. "It may not be human joy. But it is joy."

"You have to know *everyone*. Yes!" Katito Sayialel is saying, her lilting accent as clear and light as this African morning. A native Maasai, tall and capable, Katito has been studying free-living elephants with Cynthia Moss for more than two decades.

How many is "everyone"?

"I can recognize all the adult females. So," Katito considers, "nine hundred to one thousand. Say nine hundred. Yes."

Recognizing hundreds and hundreds of elephants on sight? *How* is this possible? Some she knows by marks: the position of a hole in an ear, for instance. But many, she just glances at. They're that familiar, like your friends are.

When you're studying social relationships as they're all mingling, you can't afford to say, "Wait a minute; who was *that*?" You have to know them. Knowing hundreds of individuals is necessary because elephants themselves recognize hundreds of individuals. They live in vast social networks of families and friendships and other relationships. That's why they're famous for their memory. They certainly recognize Katito.

"When I first arrived here," Katito recalls, "they heard my voice and knew I was a new person. They came to smell me. Now they know me."

Vicki Fishlock is here, too. A blue-eyed Brit in her early thirties, Vicki studied gorillas and elephants in the Republic of the Congo before bringing her doctoral diploma here to work with Cynthia. She's been here for a couple of years and has no plans to go anywhere else if she can help it. Usually Katito takes attendance and rolls on. Vicki stays and watches behavior. Today we're out on a bit of a jaunt, as they're kindly showing me around.

We come upon some grazing elephants trailing a train of egrets and an orbiting galaxy of swirling swallows. The birds rely on elephants to stir up insects as, like great gray ships, they plow through the grassy sea. Light shifts on their wide, rolling backs like sun on ocean waves. Sounds of ripping, chewing. Flap of ear. Plop of dung. The buzz of flies and swoosh of swatting tails. Soft tom-tom footfalls. And, mostly, the quiet ways of ample beasts. Wordlessly they speak of a time before human breath. They get on with their lives, ignoring us.

"They're not ignoring us," Vicki corrects. "They have an expectation of politeness, and we're fulfilling it. So they're not paying us any mind." Through hummocks and the bush, in our vehicle we amble with them.

"Here's someone feeling a little silly," Vicki says, pointing. "See her with that loose walk and her trunk swaying?"

I *do.*

"One day when I was new here," Vicki recalls, "Norah and I were watching and suddenly everyone started running around and trumpeting. I was like, 'What the hell just happened?' Norah said, 'Oh they're just being silly.'

"I thought, '*Silly*?' And the next thing I know, a full-grown female comes along walking on her knees and throwing her head around, acting just daffy. They were just happy. They were like, 'Yaaay!' Everyone says how smart they are. But they can be ridiculous, too. If a young male doesn't have a friend around, sometimes he'll make a little mock charge at us, then back up or twirl around. I actually had one male kneel down right in front of the car and throw zebra bones at me, trying to get me to play with him."

When the science of animal behavior was getting established, there was no scientific way to approach the prospect of animal emotions or to pose a question such as "What does an elephant feel when she nurses her baby?" There was nothing to go on. No one had watched free-living animals living their real lives. Brain science was in its infancy. So speculation about their feelings could only draw on our own feelings—leading ourselves in circles. The new scientists insisted on observation, not speculation. Speculation was messy guessing that one had to avoid. We can observe *what* an elephant does. There's no way to know *how* the animal feels. So just count how many minutes she nurses her offspring. As even the noted elephant communication expert Joyce Poole has explained, "I was trained to view nonhuman animals as behaving in ways that don't necessarily involve any conscious thinking."

My own initiation into formal training included the classic directive to steer strictly clear of anything smacking of attributing human mental experiences—values, thoughts, or emotions—to other animals. (Doing so is called "anthropomorphism.") I appreciate that. We shouldn't assume that animals (or, for that matter, lovers, spouses, kids, or parents) "must be" thinking and feeling just as we would if we were them. They're not us. By not assuming, we open a clearer path to understanding what's really going on.

But it wasn't that the question of animal thoughts and emotions awaited better data; it was that the whole subject became verboten. Wondering what feelings or thoughts might motivate behavioral acts became totally taboo. Radio blackout. Professional behaviorists could describe what they saw, period. Description—and *only* description—became "the" science of animal behavior. You could say that a lion was stalking a zebra. If you said the lion *wanted* to catch it, you'd be accused of "projecting your human emotions." After all, the lion might be an utterly unconscious machine—you can't know. You could say, "The elephant positioned herself between her calf and the hyena." The mother wouldn't position herself between her baby and an antelope. She knows hyenas are a threat. But if you said, "The mother positioned herself to protect her baby from the hyena," *that* was out of bounds; it was anthropomorphic. We can't know the mother's intent. And this was stifling.

In establishing the study of behavior as a science, it had originally been helpful to make *anthropomorphism* a word that raised a red flag. But as lesser intellects followed the Nobel Prize–winning pioneers, *anthropomorphism* became a pirate flag. If the word was hoisted, an attack was imminent. You wouldn't get your work published. And in the academic realm of publish or perish, jobs were at stake. Even the most informed, insightful, logical inferences about other animals' motivations, emotions, and awareness could wreck your professional prospects.

But what *is* a "human" emotion? When someone says you can't attribute human emotions to animals, they forget the key leveling detail: humans *are* animals. Human sensations *are* animal sensations. Inherited sensations, using inherited nervous systems.

All of the emotions we know of just happen to be emotions that humans feel. So, simply deciding that other animals can't have any emotions that humans feel is a cheap way to get a monopoly on all the world's feelings and motivation. People who've systematically watched or known animals realize the absurdity of this. But many others still don't. "The dilemma remains," wrote author Caitrin Nicol recently, "how to get an accurate understanding of the animals' nature and (if appropriate) emotions, without imposing on them assumptions born of a distinctly human understanding of the world."

But tell me, what "distinctly human understanding" hampers our understanding of other animals' emotions? Is it our sense of pleasure, pain, sexuality, hunger, frustration, self-preservation, defense, parental protection? We never seem to doubt that an animal acting hungry feels hungry. What reason is there to disbelieve that an elephant who seems happy is happy? We can't really claim scientific objectivity when we recognize hunger and thirst while animals are eating and drinking, exhaustion when they tire, but deny them joy and happiness as they're playing with their children and their families. Yet the science of animal behavior has long operated with that bias—and that's unscientific. In science, the simplest interpretation of evidence is often the best. When animals seem joyous in joyful contexts, joy is the simplest interpretation of the evidence. Their brains are similar to ours, they make the same hormones involved in human emotions—and that's evidence too.

The ability—and the need—to form deep social bonds developed through deep time. It didn't just suddenly appear with the emergence of modern humans. Parental care, satisfaction, friendship, compassion, grief—all began their journey in prehuman beings. Our brain's provenance is inseparable from other species' brains in the long cauldron of living time. And thus, so is our mind. Our mind arrived with other species' minds in one long gesture in the continuous sweep of Life.

Up ahead, two groups are converging, each part of the family called the FBs. All mothers are keeping in physical contact with their babies by touching them with their tails. Right now Felicity is with her daughters and two unrelated females—Flame and Flossie, who are sisters. Fanny is leading her young ones, her niece Feretia and her great-niece Felica. Vicki tells me that Fanny is very level but not hugely affectionate with her young ones. By contrast, Felicity and her offspring are always touching one another.

Felicity knows that this area they've just covered is safe. Her family feels secure at the moment because Felicity's got their back. Often a matriarch will lead from the rear of the group. But when she stops, everyone stops. They're listening to her even when she's behind them. They know right where she is. When something scary happens up ahead, the family will rush back to Felicity. If there's something dangerous, like a lion or buffalo, she may choose to retreat or have the family charge and drive them off.

"That decision is up to her," Vicki tells me. Right now, Vicki observes, "Everyone feels safe and secure, everyone's relaxed; kids are playing. Nobody's worried about anything.

"So, Felicity's an unusually good matriarch. If you have a matriarch who's a suspicious, high-stress type, everybody's always being vigilant, always listening for danger. Elephants like that continually have elevated levels of the stress hormone cortisol in their blood; that is not good for metabolism."

Vicki says to the elephants, "So it pays to be chilled out, doesn't it, guys?"

Everyone assents by calmly doing what they've been doing.

Felicity's baby is about fifty yards away from her mother, up here near us with the rest of the family. She's a particularly confident little elephant. Her big sister is right next to her. Suddenly, she runs back to her mother.

"It's a bit of a game," Vicki interprets. "Like, 'Look, I'm over here, and I'm okay!'" She's having fun, ears out, waving her little trunk around, charging an egret. It looks like the *kind* of charge an adult might use to scare off a lion. Part of the family's role is allowing youngsters to explore and learn through their own experiences. Male youngsters tend to play pushing contests against each other. Females tend to play, "I'm chasing enemies."

Felicity's baby charges a couple more egrets. "But you also have to teach them to respond to danger."

Even full-sized adults sometimes play games against imaginary enemies. They might start running through tall grass, thrashing it, the kind of behavior they might actually use to chase away lions. "But the elephants are playing," declares Vicki. "They know there are no lions."

But—if elephants act like there are lions and there are no lions, isn't it possible they're just making a mistake or being extra careful?

"It's easy to tell," Vicki explains. A serious elephant faced with a real threat pays steady attention. Playing elephants run in a loose and "floppy" way, shaking their heads to let their ears and trunk flap and flop around. "They are not making mistakes or giving false alarms. They're running around as if highly alarmed but doing what we call 'play-trumpeting.' They all know they're playing."

When doing serious things in nonserious moments—staring over their tusks at imagined enemies in the wide-eyed display or shaking their heads before charging and running away in mock fright—playing elephants often seem to be going just for the humor value. And they're all in on the game. Such blatant silliness must be—I am guessing—as close to hilarious as an elephant perceives; the elephants must be cracking themselves up. Clearly, they're having fun. "Sometimes they put bushes on their heads and just look at you like that," Vicki says. "Ridiculous."

Fanny's little one flares her ears at us, sizing us up, deciding whether we're now the enemy. She pulls herself up to full height and kind of looks down her nose at us. "We call that posture 'stand-tall,'" Vicki explains. The little one seems to decide we're either okay or too big to mess with. In a few moments she's under her sister's chin, deciding whether to charge a grouselike bird called a yellow-necked francolin.

The scene is so moving, so filled with beautiful innocence. But their lives are not always this perfect. No lives are.

Flanna's ear has a big triangle missing, where a spear went through it. One of these elephants lacks a tail. Hyenas sometimes bite off an elephant's tail while she is giving birth. Hyenas will also seize a baby if they can. Lions can kill smaller elephants. The joys and the dangers are both very real, and these babies, running around just having fun, are as naive as they are vulnerable. They have to be *taught* to fear lions.

Felicity has been leading from the rear but has slowed and dropped even farther behind, as if something is up. Suddenly she wheels, and a hyena peeks out from behind a bush. Felicity stares. Its cover blown, the hyena saunters off.

"So, see—" says Vicki rather proudly. "Felicity is *such* a good matriarch."

One day, Katito saw an elephant walking with a spear stuck in her. She went for help. Returning with a veterinarian who'd come to administer a dart filled with antibiotics and painkillers, they saw that another elephant was with the wounded one—and that the wounded one no longer had the spear in her. No one had ever heard of an elephant removing a spear from another elephant; it must have fallen out. But when the veterinarian's dart hit the wounded elephant, the friend moved in and pulled out the dart. Researchers once saw an elephant pluck up some food and place it into the mouth of another whose trunk was badly injured. "Elephants show empathy," Amboseli researchers Richard Byrne and Lucy Bates state plainly. This should come as no surprise. They aid the ailing. They *help* one another.

More mysteriously, elephants sometimes help people. George Adamson, who helped raise the famous lion Elsa of the book *Born Free*, knew an elderly, half-blind Turkana woman who'd wandered off a path; nightfall caused her to lie down under a tree. She woke in the middle of the night to see an elephant towering over her, sniffing up and down with its trunk. She was paralyzed by fear. Other elephants gathered, and they soon began breaking branches and covering her. The next morning, her faint cries attracted a herder, who released her from the cage of branches. Had the elephants mistaken her for dead and attempted to bury her? That would have been strange enough. Had they sensed her helplessness and, in empathy and perhaps even compassion, enclosed her in protection from hyenas and leopards? That would have been stranger still.

Cynthia told me of a wonderful matriarch named Big Tuskless. She died of natural causes, and a few weeks later Cynthia brought her jawbone to the research camp to determine her age at death. A few days after that, her family passed through the camp. There are several dozen elephant jaws on the ground in the camp, but the family detoured right to hers. They spent some time with it. They all touched it. And then all moved on, except one. After the others left, one stayed a long time, stroking Big Tuskless's jaw with his trunk, fondling it, turning it. He was Butch, Big Tuskless's seven-year-old son.

Do elephants grieve? And can we really know? After a young elephant dies, its mother sometimes acts depressed for many days, slowly trailing far behind her family. When a female named Tonie gave birth to a stillborn baby, she stayed with her dead child for four days, alone in the heat, guarding it from the lions who wanted it.

Grief isn't solely about life or death; it's mostly about loss of companionship. Sometimes people we know die but we don't grieve. Sometimes people we love decide to walk out of our lives, and though they remain alive, we grieve. We simply terribly miss them. Knowing them changed our lives, and losing them changes our lives. Barbara J. King says that when two or more animals have shared a life, "Grief results from love lost."

Is *love* really the right word? If an elephant sees her sister and calls to maintain contact, or a parrot sees its mate and wants to be nearer, some *feeling* of the bond makes it seek closeness. One word we use for the feeling behind our desire for closeness is *love*. Elephants and birds don't feel their love for one another the way I feel my love, but the same is true of my own friends, my mother, my wife, my step-daughter, and my next-door neighbors. Love isn't one thing, and human love isn't all identical in quality or intensity. But I believe that the word that labels ours also labels theirs. Love, as they say, is many splendored. *Love* probably *is* the right word.

A Concept of Death

CYNTHIA MOSS AND JOYCE POOLE

That elephants appear to possess "some concept of death," researcher Cynthia Moss writes in *Elephant Memories: Thirteen Years in the Life of an Elephant Family*, "is probably the single strangest thing about them." No, they do not lie down to die in special places or elephant graveyards, as one ancient myth would have it, but there seems little question that they are simultaneously fascinated and disturbed by the discovery of a dead elephant. They ordinarily act completely incurious about the remains of other species, but coming across the carcass of an elephant, they will "stop and become quiet and yet tense in a different way from anything I have seen in other situations." They react similarly to old elephant bones: examining them carefully, sometimes removing and transporting them to another place close to a familiar trail or watering spot. These animals are supremely sensitive to the chemistry of olfaction, and it is clear that they can distinguish the smell of elephant bones from those of other animals. They seem to be especially fascinated by the skulls and tusks, according to Moss in the first excerpt, and it is possible that they recognize the remains of important or beloved individuals by the distinctive smell of their bones and tusks. In the second excerpt of this chapter, from *Coming of Age with Elephants*, researcher Joyce Poole expands on the subject beautifully.

Elephants may not have a graveyard but they seem to have some concept of death. It is probably the single strangest thing about them. Unlike other animals, elephants recognize one of their own carcasses or skeletons. Although they pay no attention to the remains of other species, they always react to the body of a dead elephant. I have been with elephant families many times when this has happened. When they come upon an elephant carcass they stop and become quiet and yet tense in a different way from anything I have seen in other situations. First they reach their trunks toward the body to smell it, and then they approach slowly and cautiously and begin to touch the bones, sometimes lifting them and turning them with their feet and trunks. They seem particularly interested in the head and tusks. They run their trunk tips along the tusks and lower jaw and feel in all the crevices and hollows in the skull. I would guess they are trying to recognize the individual.

On one occasion I came upon the carcass of a young female who had been ill for many weeks. Just as I found her, the EB family, led by Echo, came into the same clearing. They stopped, became tense and very quiet, and then nervously approached. They smelled and felt the carcass and began to kick at the ground around it, digging up the dirt and putting it on the body. A few others broke off branches and palm fronds and brought them back and placed them on the carcass. At that point the warden circled overhead and dived down in his plane to guide the rangers on the ground to the dead elephant so that they could recover the tusks. The EBs were frightened by the plane and ran off. I think if they had not been disturbed they would have nearly buried the body.

Even bare, bleached old elephant bones will stop a group if they have not seen them before. It is so predictable that filmmakers have been able to get shots of elephants inspecting skeletons by bringing the bones from one place and putting them in a new spot near an elephant pathway or a water hole. Inevitably the living elephants will feel and move the bones around, sometimes picking them up and carrying them away for quite some distance before dropping them. It is a haunting and touching sight and I have no idea why they do it.

When an elephant dies in Amboseli we let it rot for a while and then collect the lower jaw for aging. The jaws are often still smelly so we put them out in the sun on the periphery of the camp. Without fail these hold a fascination to all passing elephants. Recently one of the big adult females in the population died of natural causes and we collected her jaw after a few weeks and brought it to the camp. Three days later her family happened to be passing through camp and when they smelled the jaw they detoured from their path to inspect it. One individual stayed for a long time after the others had gone, repeatedly feeling and stroking the jaw and turning it with his foot and trunk. He was the dead elephant's seven-year-old son, her youngest calf. I felt sure that he recognized it as his mother's.

There is something eerie and deeply moving about the reaction of a group of elephants to the death of one of their own. It is their silence that is most unsettling. The only sound is the slow blowing of air out of their trunks as they investigate their dead companion. It's as if even the birds have stopped singing. Just as unsettling is the way elephants back into their dead. Although elephants use their front legs for killing, by kneeling on their victims, they have a way of walking backward and using their sensitive hind feet surprisingly delicately for waking up their babies and touching the dead. Using their toenails and the soles of the feet, they touch the body ever so gently, circling, hovering above, touching again, as if by doing so they are obtaining information that we, with our more limited senses, can never understand. Their movements are in slow motion, and then, in silence, they may cover the dead with leaves and branches. Elephants' last rites? A wake, a death watch, the calling up of the elephant spirits? Elephants perform the same rituals around elephant bones. They approach slowly and silently, and then the touching begins, slowly, as they deliberately, carefully turn a skull over and over with their trunks, touching, hovering over the long bones with their hind feet. Watching elephants with their dead always leaves me with many stirring emotions and many real questions. Perhaps our fascination, or even our fear, is that elephants possess something that we believe only humans can have: a sense of death and therefore a sense of self, a sense of their place in nature.

I used to collect the lower jaws of elephants I had known, bringing them to the camp for aging, tagging them, and placing them, one by one, in a growing semicircle around my tent. Often elephants would enter the camp at night and carry the recently collected jaws away, and I would have to retrieve them the following morning from under the palms. One night an elephant came after I had brought back a new jaw and began to break off pieces of palm fronds from my *makuti* roof. As my roof began to shake, I called out for her to stop; I couldn't understand what she was doing. In the morning I found that the jaw had disappeared, and strewn about in its place were the broken-off bits of *makuti*. Had the elephant been trying to cover the mandible?

An interesting anecdote was related to me by Simon Trevor, onetime warden and later wildlife filmmaker resident in Tsavo National Park. A woman had come to visit him wearing several ivory bracelets, which were then still fashionable and acceptable to wear. As they were standing outside his house, Eleanor, an elephant who had been orphaned and raised by Daphne Sheldrick (now renowned for her tireless work saving baby elephants left orphaned by poachers) from a baby, approached them. As Eleanor came closer, Simon said to the woman, "I think you should put your ivory behind your back."

Eleanor walked up, reached behind the woman, and, taking her hand in her trunk, carefully examined the ivory. I was intrigued by this story. Could Eleanor actually smell the ivory? And why was she so interested in it? I suggested to a fellow researcher, Barbara McKnight, who was working on elephants in Tsavo, that she might try a similar experiment if she found a piece of ivory in the park. Barbara came back from the field one day and stood near the entrance to Eleanor's *boma*, or enclosure, with the ivory hidden behind her back, where Eleanor could not see it. Eleanor walked up, reached behind Barbara's back, took the ivory into her trunk, and raised it up close to her eye to study it. Not only was Eleanor apparently able to smell something that to us has no scent, but she was interested in actually studying the piece of ivory, presumably because it had belonged to another elephant.

The response of elephants to the bones and tusks of their own species has been described and filmed many times, but it is worth recounting another one of my own experiences. Recently I was assisting a National Geographic film crew to obtain footage of elephants with elephant bones. I had often collected bones and placed them in the path of a group of elephants, in order to study their reactions, but this time was different. We were gathering the bones of Jezebel, my favorite female, and presenting them to her own family. The family approached her remains and then suddenly stopped and became silent. They neared the bones very slowly and then spent the next hour turning the skull, the jaw, and the long bones over and over. The elephants, who appeared to be in a sort of trance, neither interacted nor vocalized and seemed to focus only on the dead elephant. Jolene, Jezebel's daughter, appeared to be the most absorbed of the group. What was she thinking: This is my mother; she died in a lot of pain; life is not the same without my mother? Why would an elephant stand in silence over the bones of its relative for an hour if it were not having some thoughts, *conscious* thoughts, and perhaps memories?

As I mentioned earlier, elephants do not like the sight of blood nor the animals that cause it, and will chase lions, jackals, or vultures off a kill. But they do not then stand over a dead zebra, slowly touching its body and burying it with dirt and vegetation, nor do they turn their trunks toward or pause to stand over the bones of a wildebeest. They reserve this behavior for their own kind and, sometimes, one other species: humans. If elephants see themselves as different from other animals, is it possible that they also see humans as different from the rest of nature? And how do they measure that difference?

The Secret Language of Elephants

KATY PAYNE

Acoustic biologist Katy Payne began her career studying, with her husband, Roger Payne, the progressively changing songs of humpback whales. Their early recordings of whale communications through such sonic productions led to the best-selling album *Songs of the Humpback Whale* (1970), and the success of that record contributed to a growing sense of whales, the largest mammals of the sea, as complex and intelligent animals. In 1985, Payne began to wonder about the nature of communication among the largest mammals on land. That initial curiosity led her to spend a week with a group of Asian elephants—three adult females and their three young offspring—at the Washington Park Zoo in Portland, Oregon. It was her "first week in a matriarchy," Payne would later recall. On her flight home to Ithaca, New York, as she relaxed in her seat and casually recalled her time at the zoo, the thrumming vibrations of the plane's engine reminded her of her experience with the elephants as well as previous experiences with infrasound: sound below the level of human hearing that can sometimes be felt.

In Ithaca she met with two colleagues at Cornell University who lent her some specialized equipment capable of recording and measuring infrasound. With that equipment in hand, she returned to the elephants at the Washington Park Zoo. After several days and nights of recording sounds and sonic events while marking the time and documenting the individual elephants' movements and behaviors, Payne flew back to Ithaca. She returned the recorder and played the tapes, thereby finding a particular moment when one of the females, Rosy, seemed to be calling infrasonically with a male, Tunga, who was standing outside the elephant house in his own enclosure. None of the people there at the time heard anything out of the ordinary, and yet the two elephants had been standing face-to-face, separated from each other by a concrete wall nearly three feet thick, and apparently communicating. Payne coauthored a first report of that study, "Infrasonic Sounds of Asian Elephants," which appeared in a scientific journal. Around the same time, the *New York Times* reported their work more dramatically in the article "Secret Language Found in Elephants."

Later on Payne conducted an extended series of experiments in Africa that demonstrated convincingly the reality of long-distance infrasonic communication among elephants, showing them capable of receiving infrasound from a source four kilometers away, which means that the infrasound calls could cover at least fifty square kilometers. Under ideal atmospheric conditions, so it was later calculated, such calls might travel for up to 9.8 kilometers (more than six miles) and be heard by elephants within an area of around 300 square kilometers. The following excerpt from Payne's *Silent Thunder: In the Presence of Elephants* (1998) describes her initial observations of the elephants at Washington Park Zoo in Portand.

─── ── ── ── ── ── ── ── ───

"Meet Rosy, the matriarch," says Jay Haight, from inside a cage in the elephant house in the Washington Park Zoo in Portland, Oregon. The Asian elephant looms above him as they walk toward the thick vertical bars that separate them from me. Jay's shoulder is level with the bottom of Rosy's right ear, which, like her forehead and trunk, is pink-bordered with delicate pale tan and gray spots. The spots are denser at the bottom than at the top, as if they were particles slowly sliding down a liquid poured from the top of her forehead.

Her body rises and sinks hugely with each step. One shoulder at a time shifts upward and bulges; one knee at a time straightens and accepts weight; underneath the vast belly the opposite leg moves forward; the broad toe-nailed foot swings forward just above the concrete floor, sets itself down, and splays out. Steadily, the feet, legs, and shoulders shift, muscles alternating, bulk flowing forward, huge and slow. The eyes are looking down; the speckled ears are waving slowly and symmetrically in and out; the face—well, to call it a face . . .

"A-l-l right," says Jay, and the feet and shoulders and belly and back come to a rest, sinking a little. I look up at an immense gray forehead. At its base the forehead gives way to the broad top of the freckled trunk, larger in girth than my girth, longer than I am tall, on each side a gray cheek wider than my torso. From a wrinkled leather pocket on the forward side of the left cheek a patient amber eye looks down. I can see that eye but not the other. Raising myself on tiptoes and leaning close to pat Rosy's forehead, I can't see either one. Her mouth is hidden, too, under her trunk. The dry soft leather of her forehead is warm, warmer than my hand. I slide my hand down the wall of the forehead to the top of the trunk and then down the trunk, passing it over row upon row of thick warm wrinkles. All the time the trunk is moving, its tip searching and reaching this way and that, hovering, whiffing, eventually reaching tentatively through the bars and approaching my other hand, which I hold open and still.

"Hello, Rosy," I say in a low voice, and under the hand that is still stroking the trunk I feel a shiver.

"A-l-l right," says Jay to Rosy, and to me, "Meet her son Rama. Thirteen months old. HEY there, Rama, watch out!"

Out of the top of Rama's head and along his back spring wiry orange and black hairs. He looks up from below me with a full face, two wild eyes visible at once. The whites of the eyes show; he looks surprised, and I respond with a smile. Pressing against his mother, Rama stretches his short, stubby trunk through the bars next to Rosy's long, supple one, and the open tip of each is pink and flexible, cool and wet as it gropes and sniffs my hand.

"This one is Hanako. GET OVER HERE, HANAKO! BACK UP, ROSE. BACK, RAMA! BACK!! Hanako's big boy here is our first grandchild. He's nineteen months old. Git over, Look-Chai, meet Katy."

"Ah!" I hold out my hand, but it is my sandaled feet that the elephants' trunks are delicately exploring, tickling them with breath.

A tall, quiet keeper named Jim Sanford now joins Jay in the cage: he's brought a wheelbarrow and two shovels. The two men fill the barrow with elephant dung. The elephants seem glad that Jim has come: I notice the gladness as a relaxation, and I feel a faint thrill in the air and hear a gentle rumble as he strokes Rosy's trunk between her eyes. The men stand together, each rubbing a different elephant. The elephants' trunks reach around the men's bodies, sniffing.

The keepers ask me what inspired me to come for this visit. I tell them I'm an acoustic biologist from Cornell University, and I've been wondering what kinds of sounds elephants make. I've spent the last fifteen years studying the songs of whales, which are long and complex, and change continuously and progressively. Last week some colleagues who also study culture (learned behavior) in animals invited me to California so we could compare our findings. That brought me within reach of this zoo, with its eleven elephants. I called Warren Iliff, the zoo's director, and he said sure, come ahead, you can spend the first week in May with our elephants if you like.

Jim and Jay tell me about Look-Chai's heritage and the circumstances of his birth. His grandmother, old Tuy Hoa from Vietnam, was the zoo's first elephant. She gave birth to Hanako in 1963. Hanako grew up in the zoo and in her nineteen years gave birth four times; but of her calves only Look-Chai has survived.

Tuy Hoa was old and arthritic during Hanako's fourth pregnancy. The vet decided not to risk having her present at the birth, for even standing up was harmful to her. But they moved Hanako into an adjacent cage where the mother and daughter would be able to smell and hear each other, for a grandmother elephant would naturally assist in birthing. Hanako labored for two days without giving birth. At last, exhausted and worried, the vet and keepers opened the gate connecting the cages. As Hanako ran to her mother both elephants bellowed, rumbled, trumpeted, and screamed, and from other elephant cages farther back in the building came answering rumbles, trumpets, and screams. Then Hanako dropped Look-Chai onto the floor beside his grandmother.

Within two weeks an extraordinary thing happened. Milk formed in the old grandmother's breasts, although she had not had a calf of her own for several years. Along with Hanako, Tuy Hoa nursed the last of her descendants to be born in her lifetime.

"This one is Pet," says Jay. "GET BACK, LOOK-CHAI! Pet's the bottom of the social heap. She'll do anything to stay subordinate—you'll see." Pet's yearling daughter, Sunshine, is pushing forward to join her, and now six trunks are hovering and gliding a few inches from my feet, legs, hands, and belly.

Suddenly Jay makes a decision: no more trunks allowed on my side of the bars. He shouts a volley of commands, and swats each reaching trunk with the flat side of an elephant hook until they reluctantly withdraw.

"And don't you let them sniff you if we're not here," says Jay to me. Rama starts to break the rule. "No, Rama," I say.

"But *you* should not discipline them," says Jay, quickly. "GET BACK, RAMA! The only people who should try to establish discipline are the ones who will enforce it. We'll tell you stories..." And they did. Only a fool attempts to read the mind of an elephant, and I heard about quantities of fools. Each of them had been a fool at least once. Fools had been kneeled on and pressed against concrete walls; some had died. "You'll be noodles if they drag you through the bars," the men warned me. They finished cleaning the cage, ordered the elephants to stay away from me, and left for lunch.

The indoor air was chilly and dank, and I was tired. I leaned my head against one of the bars, thinking how to begin, and closed my eyes. But a sudden feeling of warmth on my left shoulder caused me to open them again. The radiator was Rosy's body—she had moved up against the bars as close to me as she could get, beginning a process that was repeated each time the keepers left, a slow, gradual migration of all the elephants in my direction. Six trunks reached slightly through the bars, gently surrounding me with whiffing as the elephants decided, more deliberately than before, who I was.

Thus began my first week in a matriarchy. A calm, pervasive discipline regulated the elephants' behavior, for the three adults, born on separate continents and thrown together in captivity, had established a dynamic order among themselves. Rosy, though smaller than Hanako, was the eldest, and whenever no keepers were present her authority was supreme.

It was an authority that extended to and included me. I thought about that as I listened to the whiffing, the inhaling of my smell along with the smells of one another's dung and urine and flesh along with the hay and fruit and nuts on the cage floor. Of what did the authority consist, and how was it communicated among the elephants? I sensed a comprehensive but relaxed mutual attention. Rosy was granting her herd the privilege of exploring my smell carefully. She was granting me the privilege of being carefully smelled. She growled when a calf became too inquisitive, and the little trunk hastily withdrew. Not sure how to keep my end of the bargain, I stayed quiet and kept my hands still, without hiding my interest. When an adult moved in close to the bars I looked up into the hazy dark pupil of her solemn eye. The pupil was so large that I never felt our eyes really met: I didn't know whether I was in or out of focus. I wished that I had a hovering and whiffing trunk of my own so I could learn the same things about the elephants that they were learning about me.

The square, burly babies moved on to nurse and explore with their supple trunks. Scooping up, sucking in, puffing, sniffing, blowing out, they took in every detail of the floors and walls. In a burst of vertical curiosity all trunks lifted at once, and for a minute or two the three little elephants walked about exploring smells high over their heads. Encountering one another, they entwined trunks, putting the tips in one another's mouths or ears and sniffing. They invented strategy upon strategy for picking things up, with mixed success. Their fuzzy heads were littered with bits of slung hay. They groomed one another and put what they found in their mouths or ears, or poofed the scraps into the corners of the stall. They galloped stiff-legged, heads high, trunks surging, feet loose and floppy, foreheads wrinkled up and eyes wide. At a certain level of abandon, the mothers subdued and separated them. In this they reminded me of the mother right whales that Roger, my husband, and I had watched a decade earlier from the cliffs of the Peninsula Valdés in Argentina. When the whales subdued and separated their calves, we had surmised that the mothers were saving their collective metabolic energy for the long migration across the South Atlantic Ocean to South Georgia and back.

In the back room, our feet up on desks, their hooks laid down and my pencil taken up, the keepers and I compared experiences. They wanted to know what whales were like. I told them about watching the sea hour after hour through binoculars, searching for distant dark shapes—teardrops and exclamation points—which were the caps of whales' backs visible on the surface on a clear and calm day. There were lucky hours when a group of whales would come so close under our cliff that we could identify them as individuals; then for the rest of the day we'd watch them slide away across the silver sea to line up parallel and slip down under, perhaps to rise again a half mile farther out, and be joined by others, and line up parallel, and disappear again, heading for the mouth of the Golfo San José, from which if you go due east, the first land you will strike is New Zealand.

They told me about one swaying elephant belly at a time. About elephants' knees in their faces. About dangerous, affectionate, and often inscrutable individuals. The elephants they knew were as full of quirks and idiosyncrasies as an assortment of very weird people. The confinement and echoing walls of the elephant house amplified whatever propensities each captive animal and each keeper had for dominance, retribution, compassion, and caregiving. These exaggerated personalities were the basis of the relationships that developed within the walls.

Except for an occasional break to run down a path through a forest just below the zoo, and then up again, I spent the whole of every day in the elephant house. Elephants may not have been the only interesting animals in the zoo, but I had eyes, or ears, only for them. At the end of the week I boarded a plane and started the journey home. My ears and the back of my neck were itchy with pins and needles of straw and hay and other zoo frass. I tucked a plastic bag into the overhead bin—my barn jacket, bagged to preserve its smell. I sighed to think that the warm beings who had taken such an interest in *my* smell would become a fading memory as I got back to normal life.

The sigh acknowledged a slight failure as well as sadness. I would have liked to learn something new, but it seemed that the time was not ripe. I closed my eyes to review the happenings I'd witnessed in the zoo. I would glean them, and then say good-bye.

Here stood Pet in the back corner, the end of her trunk moving over the floor like a squeegee, collecting together a few last stems and scraps in the hour before feeding time. Here came Hanako and Rama sauntering in her direction. Were they coming to visit her, or to deprive her of what she had gathered? The latter, I decided, and I watched carefully, thinking of Jay's comment that Pet would do any-thing to stay on the bottom of the hierarchy. I heard a faint rumble and the animals shifted, but Pet held her own. The air thrilled a little: I felt happy for Pet.

Here came Jim with the grain and hay. All the elephants moved to greet him, and I enjoyed his gentle voice, and again sensed a kind of thrill in the air.

Here stood little Sunshine reaching toward me through the bars; behind her, her mother standing by, to "turn me into noodles" should I prove untrustworthy. The airplane throbbed, reminding me of the faint throbbing, or thrilling, or shuddering I'd felt at that moment. It had been like the feeling of thunder but there'd been no thunder. There had been no loud sound at all, just throbbing and then nothing.

Now a recollection from more than thirty years earlier joined the first. I was thirteen years old, and I was standing not in a zoo but in Sage Chapel at Cornell University in Ithaca, New York. And what I was hearing was not silence but enormous chords from a pipe organ that was accompanying singers, and I was one of the singers. My mother was across the room in the alto section. The organist and conductor were lit by lamps that illuminated their music. The little circles of yellow light from the lamps were framed by the darkness of early evening inside a vast building. High up in the space a series of round stained-glass windows, still receiving light from the sky, glowed down on us.

The organ was alive. In a powerful combination of voices it was introducing the great chorus that opens the second half of Bach's *Passion According to St. Matthew*; we were drawing breath to sing, "Oh man, bewail thy grievous sin." The organist pulled out the great stop and the air around me began to shudder and throb. The bass notes descended in a scale. The deeper they went, the slower the shuddering became. The pitch grew indistinct and muffled, yet the shuddering got stronger. I felt what I could not hear. My ears were approaching the lower limit of their ability to perceive vibrations as sound.

Is that what I was feeling as I sat beside the elephant cage? Sound too low for me to hear, yet so powerful it caused the air to throb? Were the elephants calling to each other in infrasound?

Earthquakes, volcanic eruptions, wind, thunder, and ocean storms—gigantic motions of earth, air, fire, and water—these are the main sources of infrasound, sound below the range of human hearing, which travels huge distances though rock, water, and air. Among animals only the great fin and blue whales were known to make powerful infrasonic calls. No land animal approached the mass or power of these great mammals of the sea, but now I wondered: might elephants, too, be using infrasound in communication?

Elephant in the Mirror

JOSHUA M. PLOTNIK, FRANS B. M. DE WAAL,
AND DIANA REISS

In 1970, American psychologist Gordon G. Gallup Jr. described a series of experiments that seemed to demonstrate a decisive cognitive
distinction between great apes and monkeys. Gallup began his experiments with four preadolescent chimpanzees (members of a great ape
species) who were each isolated within a cage inside a bare room and
presented for several days with a large mirror to consider. Researchers
observed the apes in secret and recorded the animals' behavioral responses to their own reflections in the mirror.

None of the four chimpanzees had seen a mirror before, so it is
unsurprising that their initial response seemed confused or naive. They
behaved in a social manner toward the image in the mirror, as if they
were seeing another chimp rather than their own reflection: responding to the image with threats, head-bobbing, vocalizations, and so on.
Within a few days, however, the social responses declined markedly
and were replaced by self-referential responses, as if the animals finally
understood that they were seeing themselves. They began using the
mirror to look at parts of their bodies they couldn't otherwise see and
to watch themselves manipulating wads of food in their mouths, picking pieces of food from their teeth, picking their noses, contorting their
faces, blowing bubbles.

In a second phase of the experiment, Gallup arranged for each of the
four chimpanzees to be knocked out with an anesthetic. While unconscious, they were each marked on an ear and the forehead with spots

of an odorless red dye. After the animals had recovered fully from the anesthesia, each one was individually observed for thirty minutes without the mirror present to register the number of times they touched the spot. A single chimp spontaneously touched the facial spot without the presence of a mirror. After the mirror was returned, all four chimps touched the spot more than twenty-five times during a half hour of observation. Gallup concluded that the chimpanzees had demonstrated both a sense of self and the "capacity for self-recognition."[46]

Later iterations of Gallup's mirror test have considerably expanded our concept of which species are capable of learning to recognize themselves in a mirror and which are not. We now accept that members of at least four of the five great apes (humans, chimpanzees, bonobos, orangutans, and—the possible exception—gorillas) will under ordinary circumstances pass the mirror test. We also accept that no species of monkeys will pass the test. Monkeys and apes are both groups within the primate order, and as primates they share many features. Perhaps the most obvious difference between those two major primate groups is that of size: apes are generally bigger and have significantly larger brains than monkeys. If we consider this feature alone, it becomes clear that elephants ought to be another candidate for Gallup's mirror test.[47]

The 2006 report excerpted below describes the first scientific demonstration that Asian elephants—like humans, great apes, and individuals from a few other exceptional species (bottlenose dolphins, killer whales, and Eurasian magpies)—will recognize themselves in a mirror.

———— ━━ ━ ━ ━━ ━ ━━ ━━ ━ ━ ————

Mirror self-recognition (MSR) is exceedingly rare in the animal kingdom. Attempts to demonstrate MSR outside of the Hominoidea (humans and apes) have thus far failed, with the notable exception of one report on dolphins. Animals that demonstrate

MSR typically go through four stages: (1) social response, (2) physical mirror inspection (e.g., looking behind the mirror), (3) repetitive mirror-testing (the beginning of mirror understanding), and (4) self-directed behavior (the recognition of the mirror image as self). The final stage is verified if a subject passed the "mark test" by spontaneously using the mirror to touch an otherwise imperceptible mark on its own body. Application of the mark is recommended only if the preceding criteria have been met. Animals without MSR tend to remain at stages 1 and 2. Even if the degree to which the mirror image is confused with a stranger is debatable for some non-MSR species, these animals likely lack understanding of who is in the mirror.

Gallup was the first to hypothesize about a phylogenetic connection between MSR and empathy, a connection supported by evidence for consolation behavior in apes but not monkeys. A possible ontogenetic connection between MSR and empathy is reflected in the coemergence of MSR and "sympathetic concern" during child development. Dolphins and elephants represent interesting additions to MSR tests because, like the hominoids, they are highly empathic animals known for so-called "targeted" helping (i.e., helping that takes the needs of others into account) aimed at both conspecifics and humans. As in dolphins, there are numerous reports of elephants physically supporting or trying to lift up injured or incapacitated conspecifics. In view of the aforementioned hypothetical connection with empathy, the elephant's known social complexity and its relatively large and complex brain, we introduced three adult female Asian elephants—Happy, Maxine, and Patty—at the Bronx Zoo in New York City to a jumbo-size mirror (244 × 244 centimeters) in a variant of the classical mark test using both visual and "sham" marks.

Elephants have the advantage that they can touch most of their own body with their trunks, thus permitting an unequivocal mark test. A previous failed attempt to demonstrate MSR in two Asian elephants presented the animals with a relatively small mirror that was kept at a distance, well out of trunk-reach. Assuming that physical exploration of the mirror should be part of the learning process and that mirror size matters, we built an almost 2.5-meter-tall elephant-resistant mirror to allow close-up inspection of the reflective surface. Here we demonstrate that all three subjects reached the aforementioned third and fourth stages of MSR progression and that one subject passed the mark test.

Results and Discussion

There were five experimental phases: baseline (T1), covered mirror (T2), open mirror (T3), covered-mirror sham (T4), and the mark test (T5). Happy, Maxine, and Patty all spent far more time close to the mirror during three days of open vs. covered mirror (i.e., T3 vs. T2), indicating that time spent at the mirror was due to its reflective qualities rather than the novelty of the apparatus.

During T3, all three subjects showed investigative behavior of the mirror surface and frame including touching and probable sniffing. For Maxine and Patty, trunk-over-wall exploration (i.e., the swinging of the trunk over and behind the wall on which the mirror was mounted) declined from the first through the fourth day of mirror exposure. Happy never put her trunk over the mirror wall. Maxine and Patty also attempted to physically climb the mirror wall to look over and behind it, and both, on separate occasions, seemed to try to get their trunks underneath and behind the mirror by kneeling down in front of it. Their behavior was highly unusual (animal-care staff had rarely observed similar attempts by the elephants to look over or underneath enclosure walls). Remarkably all three subjects showed a total absence of social interaction with their mirror image, such as species-typical visual, vocal, or agonistic displays.

All three elephants displayed behavior consistent with mirror-testing and self-directed behavior during T3 (open mirror) and
T5 (mark test), such as bringing food to and eating right in front of
the mirror (a rare location for such activity), repetitive, nonstereo-
typical trunk and body movements (both vertically and horizontally)
in front of the mirror, and rhythmic head movements in and out of
mirror view; such behavior was not observed in the absence of the
mirror. On more than one occasion, the elephants stuck their trunks
into their mouths in front of the mirror or slowly and methodically
moved their trunks from the top of the mirror surface downward. In
one instance, Maxine put her trunk tip-first into her mouth at the
mirror, as if inspecting the interior of her oral cavity, and in another
instance, she used her trunk to pull her ear slowly forward toward
the mirror. Because these behaviors were never observed in T1 and
T2 (the initial, "no mirror" control conditions), or at any other time,
they indicate the elephants' tendency to use the mirror as a tool to
investigate their own bodies. Apes are known for very similar self-in-
vestigation in front of the mirror, such as picking with their fingers
at their teeth, which is considered a precondition for the mark test.
Similar to the time frame observed in chimpanzees, Happy reached
the criterion for the mark test within three days, and Maxine and
Patty reached this criterion within four days.

On the first day of the mark test (T5), a visible X-shaped white mark
was applied to the right side of each elephant's head, and an invisible
sham-mark was applied to the left side of the head. The sham-mark
controlled for both olfactory and tactile cues (for example, texture),
leaving only a visual component to differentiate between the mark and
the sham-mark. Lone sham-marks had been used previously in T4 to
test this control while avoiding habituation by the other elephants to
the visual component of the mark. The elephants never touched the
sham-mark under this previous condition, suggesting the predicted

absence of odor or tactile cues. (A controlled-mark condition similar to T4, but using the visual mark instead of the sham-mark in a covered-mirror condition, would have been an ideal addition to our testing procedure but could not be implemented because of the presence of the elephant's partner. Husbandry concerns prevented us from isolating each elephant during testing; hence any visual mark might have attracted the partner's attention and risked the loss of mark salience by the time actual mark tests were conducted in front of the open mirror.)

One elephant, Happy, passed the mark test on the first day of marking. Caretakers did not notice her touching either the mark or sham-mark before being released into the elephant yard. After being released into the yard, she walked straight to the mirror, where she spent ten seconds, then walked away. Seven minutes later she returned to the mirror, and over the course of the next minute she moved in and out of view of the mirror a couple of times, until she moved away again. In the following ninety seconds, out of the view of the mirror, she repeatedly touched the visible mark but not the sham-mark. She then returned to the mirror, and while standing directly in front of it, repeatedly touched and further investigated the visible mark with her trunk.

On the mark day itself, Happy showed dramatically increased head touching early in the session, most of which (twelve out of fourteen times) occurred during or within ninety seconds after proximity to the mirror. All twelve touches during or right after mirror exposure came in contact with or close to (within twenty centimeters) the visible mark on the right side of Happy's head. Happy's right-side vs. left-side bias on the first mark day differed significantly from that for head touches on nonmark days. In other words, Happy's touching of the right side of her head (particularly the mark itself) on the first mark day deviated from her general head touching during all previous conditions in both its higher frequency and its bias toward the side with the visible mark.

In contrast to Happy, Maxine and Patty failed to show increased self-touching of either the mark or sham-mark. Maxine and Patty were marked twice; Happy was marked three times. On the second day of marking, Happy never approached the mirror nor did she touch either the mark or the sham-mark. She was re-marked on a third day during which time she stationed herself at the mirror but touched neither the mark nor the sham-mark. We repeated the marking procedure on all three elephants after two months; none of them touched either the mark or sham-mark during this second phase of marking, although all of them continued to show self-directed behavior toward the mirror.

The fact that one elephant passed the mark test but two did not is not inconsistent with data recorded for other species because even in the most extensively tested species with MSR, the chimpanzee, fewer than half of the individuals may pass the test according to some studies. Similarly, absence of responses to the mark after multiple exposures resembles the reaction of MSG-capable apes, which generally lose interest in the mark within minutes of mirror exposure, apparently realizing that the mark is inconsequential. For this reason, multiple mark tests on a single individual are considered to compromise mark salience and are therefore uninterpretable.

Happy, Maxine, and Patty all continued to show self-directed behavior at the mirror, indicating that they may have only lost interest in the mark but not in their own reflection. Although we encourage further testing of Asian elephants to the marking materials and the descriptions of elephants' extraordinary memory, we would not expect these three elephants to pass during further rounds of testing.

According to the manufacturer, the ingredients of the mark and sham-mark material (both face paints) are identical except for the pigmentation components; the titanium dioxide, which is used to make the mark paint white, is odorless. The zinc sulfide, which is used to make the sham-mark paint luminescent, may have a slight odor, but we would expect that if the zinc sulfide odor was detectable and differentiated from the mark by the elephants, they would have been attracted to the sham-mark rather than the visual mark. However, as our data show, no such attraction was evident. Therefore, we conclude that the odor and tactile components of the mark and the sham-mark are either equal or negligible and that any differential touching of the mark should be due to its visible component.

The later negative outcomes in all three subjects seem to confirm the absence of tactile and odor cues of the marks. One would further suspect a lack of "concern" about bodily appearance and cleanliness in elephants compared with primates. Whereas primates often groom specific spots on their bodies, elephants rarely autogroom with their trunks. Rather, they "substrate groom," which includes dust-throwing and mud-bathing. This manner of "grooming" actually adds debris to the body. It may very well be that because of an elephant's large body-surface area and the mud and sand it often carries around on its body, attention to detail is not a priority. A small paint mark may be trivial to them.

The behavior of the elephants was strikingly similar to that of other animals who have demonstrated MSR. Although none of the elephants aimed social behavior at the mirror, they all, like the apes and dolphins, exhibited exploratory and mirror-testing behavior before more explicitly self-directed activities. Further studies on elephants of different ages, gender, and personal history will be needed to confirm and further elucidate the capacity for MSR in these animals. Our study suggests that mirror size and access to the mirror surface should be considered in replication attempts.

The mark-touching by one elephant is compelling evidence that this species has the capacity to recognize itself in a mirror. Finding strong parallels among apes, dolphins, and elephants in both the progression of behavioral stages and actual responses to a mirror provides compelling evidence for convergent cognitive evolution. Perhaps MSR indexes an increased self-other distinction that also underlies the social complexity and altruistic tendencies shared among these large-brained animals.

An Interest in Skulls and Ivory

KAREN MCCOMB, LUCY BAKER, AND CYNTHIA MOSS

Despite several intriguing anecdotes, as well as close observations at Amboseli, Manyara, and elsewhere, no one had scientifically investigated the fascination elephants are said to show for the bones and ivory of dead elephants until Karen McComb and Lucy Baker, both from the University of Sussex, collaborated with Cynthia Moss at Amboseli to develop a series of formal experiments. They wanted to test both whether the Amboseli elephants are indeed "attracted to the skulls and ivory of other elephants" and whether there was any special attraction to the remains of dead relatives.

In three different experimental events, the researchers compared elephants' interest in four kinds of objects: elephant skulls, skulls from other animals of comparable size, pieces of elephant tusk, and pieces of wood. The objects were presented in groups of three, and the researchers observed and videotaped elephants as they responded to the presence of novel objects. The researchers intended to measure the intensity of interest of test subjects—adult elephants only—in those various objects, which they measured through time spent in certain "high-interest" behaviors, specifically olfactory examination (trunk tip less than a meter from the object), probing with the trunk, or, since elephants frequently examine novel items with their feet, touching with a foot.

Each of the three events presented an array of three novel items. In the first event, an elephant skull, a piece of ivory, and a piece of wood were each placed near (twenty-five to thirty meters away from) nineteen elephant families, while the researchers videotaped the animals' reactions from inside a vehicle at a suitable distance. The results, compiled by adding up the cumulative time spent in high-interest activity for each object, showed the greatest interest in ivory—with the cumulative time of investigatory activity spent toward elephant skulls about half that of the time spent toward ivory. Time spent examining wood was far lower than either the elephant skull or the ivory.

In the second event, seventeen elephants were presented with three large animal skulls—from an elephant, a buffalo, and a rhino. The results showed a much stronger level (around twice the amount) of high-interest time spent examining the elephant skull than either of the other two skulls.

For the final experimental event, researchers presented an array of three skulls from three known matriarchs to each of the three families who had recently experienced the loss of those matriarchs. The researchers hoped to test whether individuals would recognize and favor the skull of a relative. In this case, however, the result was negative. The researchers found no significant difference between the intensity of investigation for the skulls of close relatives and those of nonrelatives. Unfortunately, a serious flaw in the research protocol makes the absence of a positive result for the third event close to meaningless as a study of elephant perceptions and preferences. The skulls chosen for these experiments were all entirely free of flesh and had been bleached by long-term exposure to the sun. In addition, the researchers washed all the test objects with a cleaning solution (Teepol) having "a low number of contaminant volatiles," then rinsed them twice and allowed them to dry before and after each experiment. The intention was primarily to "control for extraneous differences in scent between the objects prior to the experiments"—yet elephants, who are vastly more sensitive to olfactory information than we are, use scent as a primary means of gathering information about the world around them. In many circumstances, scent will be most immediate and direct way by which elephants distinguish friend from foe, stranger from relative, one object from another. Eliminating or masking olfactory cues in a test of elephant interests and perceptions, then, is equivalent to blindfolding a person before asking him or her to take a comparable test. Under these utterly unnatural constraints, any positive identification would be extraordinary impressive, while all negative ones will be meaningless.

1. Introduction

In contrast to humans, who attach great importance to the dead bodies of other humans, most mammals show only passing interest in the dead remains of their own or other species. Lions are typical in this respect, briefly sniffing or licking the dead body of a con-specific which, in the case of recently killed individuals, may subsequently be eaten. In chimpanzees (*Pan troglodytes*), interactions with dead social partners are more prolonged and complex than reported in other species, but here companions tend to leave the carcass when it starts to decompose significantly, and do not appear to interact with the bones once the carcass has rotted. In comparison, African elephants are reported not only to exhibit unusual behaviours on encountering the bodies of dead con-specifics, becoming highly agitated and investigating them with the trunk and feet, but also to pay considerable attention to the skulls, ivory and associated bones of elephants that are long dead. It has also been suggested that elephants specifically visit the bones of dead relatives. Despite being widely reported, there have been no attempts to elicit and investigate these unusual behaviours experimentally. Apparent interest in skulls and ivory may simply reflect a strong response to novel objects or a fidelity to certain routes, and the nature and specificity of elephant responses to the remains of other elephants can only be unambiguously determined using controlled experiments. Here, we use experimental presentations of object arrays to test whether elephants are indeed specifically attracted to the skulls and ivory of other elephants, and whether they show particular interest in these remains when they originate from their relatives.

2. Material and Methods

(A) STUDY POPULATION

The research was conducted in Amboseli National Park, Kenya, where long-term data on life histories have been obtained for more than 2,200 individual elephants. The primary social unit in African elephants is the female family unit, composed of adult females that are usually matrilineal relatives and their immature offspring.

(B) PROCEDURE FOR PRESENTATIONS

Between July 1998 and January 2000, free-ranging African elephants in the study population were presented with animal skulls, ivory and natural objects to investigate: (i) whether they are attracted to elephant skulls and ivory over other objects; (ii) whether they show more interest in elephant skulls than in skulls of other large terrestrial mammals; and (iii) whether they particularly select the skulls of relatives for investigation. Controlled choice tests were achieved by presenting family units with arrays of three objects in which the location of each item in line (left, center, right with respect to the approaching elephants) was systematically varied between trials to randomize the effects of preferences for particular positions.

For each presentation a suitable family unit (or section of a family unit) was identified and a set of three objects (details of different choice sets below) was decanted from the research vehicle and placed at a distance of twenty-five to thirty meters from the nearest individual in the family group. The three objects were placed in a line on the ground with one meter separating the central object from each of its neighbors. The vehicle was then driven to a position where the trial could be observed and video-recorded.

In the first experiment an elephant skull, a piece of ivory and a piece of wood were presented to nineteen different family groups, while in the second, seventeen family groups were presented with an elephant skull, a buffalo skull and a rhinoceros skull. In the third experiment, each of three families that had lost their matriarch in the recent past (last one to five years) were presented with the choice between the skull of their own matriarch and those of the matriarchs of the other two families. All three families chose between the same three skulls, with the skull that represented the matriarch for any one family representing a nonmatriarch for the other two, and each family received the choice three times, with their own matriarch's skull in each of the three possible positions in the array. The matriarch is the oldest female in the family unit, and plays an important role in coordinating the group's activities.

In the first two experiments, two different exemplars of each of the objects were used in the course of the presentations, while in experiment 3, the three objects were skulls from previous matriarchs of the JA/YA family (Jezebel), the TA family (Tuskless) and the AA family (Wartear), who had all died between one and five years before their skulls were used in choice tests. At least one week was left between different presentations to the same family. All the skulls used in the experiments were completely rotted down, so that there was no remaining flesh and the bone was bleached white by the sun. All items were washed with a solution of Teepol (which has a low number of contaminant volatiles), given two thorough rinses and air dried before and after experiments. This both controlled for any extraneous differences in scent between the objects prior to the experiments, and prevented accumulation of scent (from handling or elephant interest in particular objects) during the experiments.

(C) BEHAVIORAL RESPONSES

The elephants typically approached the objects and began investigating them by smelling and touching individual objects with their trunks and, more rarely, placing their feet lightly against particular objects and manipulating them (similar behaviors are observed during natural encounters with elephant remains). The responses of subjects to each presentation were recorded using a Sony CCD TR550E video recorder. Trials were terminated when all the individuals had finished investigating the objects and moved on, or if an individual carried an object four or more elephant lengths away from the other items. From video recordings of their responses, we calculated the cumulative amount of time that adult group members (eleven years or older) spent smelling towards an object with the trunk tip less than one meter from it, or touching the object with the trunk (high interest activity). In addition, in the case of trials involving ivory, where placing of the foot on top of objects occurred fairly regularly, the cumulative time spent by any adult touching an item with its foot was also calculated (foot on object). Elephants have mechanoreceptors in their feet, and foot placement on objects may enable them to gather tactile information.

3. Results

Subjects directed significantly different amounts of high interest activity towards the three objects in the first experiment, exhibiting a strong preference for ivory over each of the other two objects and for the elephant skull over wood (Friedman Test $N = 19$, $c^2 = 22.81$, d.f. $= 2$, $p = < 0.001$; Wilcoxon Signed Ranks Test for ivory versus elephant skull: $Z = 3.58$; $p < 0.001$, for ivory versus wood: $Z = 3.70$; $p < 0.001$, and for elephant skull versus wood $Z = 3.29$; $p < 0.005$). In the second experiment, interest in the three types of animal skull also differed, subjects exhibiting more interest in the elephant skull than in the buffalo or rhino skulls, but not in the buffalo skull compared with the rhino skull (Friedman Test $N = 17$, $c^2 = 7.12$, d.f. $= 2$, $p = < 0.05$; Wilcoxon Signed Ranks Test for elephant versus buffalo skull: $Z = 2.27$; $p < 0.05$, for elephant versus rhino skull: $Z = 2.56$; $p = 0.01$, and for buffalo skull versus rhino skull $Z = 0.25$; n.s.). In the final experiment, subjects did not direct significantly more high interest activity towards the skull of their own matriarch than towards the skulls of the two other matriarchs (binomial test on number of trials where skull that received the most attention was own matriarch's skull: $N = 9$, $k = 4$, $p = 0.35$). Due to constraints on the sample size, this experiment would be effective in demonstrating a strong preference for the correct matriarch's skull (for probabilities of success under the alternative hypothesis of 0.7, 0.8 and 0.9, power of tests would be 0.730, 0.914 and 0.992, respectively), but not a weak one (for probabilities of success under the alternative hypothesis of 0.4, 0.5, 0.6, power of tests would be 0.099, 0.254 and 0.483, respectively).

Where touching of objects with the foot was measured subjects spent significantly more time with the foot placed on ivory than on the elephant skull or wood, but not more on the elephant skull than on wood (foot on object: Friedman Test $N = 19$, $c^2 = 9.864$, d.f. $= 2$, $p = < 0.01$; Wilcoxon Signed Ranks Test for ivory versus elephant skull: $Z = 2.511$; $p < 0.05$, for ivory versus wood: $Z = 2.761$; $p < 0.01$ and for elephant skull versus wood $Z = 0.943$; n.s.).

4. Discussion

Our experiments cast light on why elephants are often observed interacting with the skulls and ivory of dead social companions—they appear to choose these items for investigation in preference to skulls from other animals or natural objects. Their preference for ivory was very marked, with ivory not only receiving excessive attention in comparison with wood but also being selected significantly more than the elephant skull. Subjects also placed their feet on or against the ivory significantly more often than on other objects. Interest in ivory may be enhanced because of its connection with living elephants, individuals sometimes touching the ivory of others with their trunks during social behavior. In experiments where no ivory was present, other items in the array appeared to receive less high interest activity overall. Despite this, the elephant skull was clearly selected for attention over the buffalo and rhinoceros skulls and over the wood. It is important to note that our findings cannot be explained by the elephants simply choosing the largest, most complex objects (the object that received the most attention overall was the ivory, which is smallest in size and simplest in shape) or the most novel ones (the rarest object was the rhinoceros skull, but this did not receive most attention).

Although there are suggestions in the literature that elephants selectively visit the bones of their relatives, our matriarch skulls presentations did not reveal a strong preference in experimental subjects for investigating the skull of their matriarch over skulls of unrelated females. While the sample size for this experiment was unavoidably limited to nine (three families presented with their matriarch's skull in each of the three positions in the array), reducing the power of the test, there was no evidence of the marked difference in interest in the three objects that was so clear in the first two experiments.

Reports of elephant graveyards, specific places where old elephants go to die, have been exposed as myths—where large concentrations of elephant bones have been found their occurrence can be adequately explained by hunting practices or mass die-offs during periods of drought. Our results suggest that elephants may not specifically select the skulls of their own relatives for investigation, but their strong interest in the ivory and skulls of their own species means that they would be highly likely to visit the bones of relatives who die within their own home range. This is the most likely explanation for why elephants have sometimes been observed interacting with the bones of particular family members, although it remains possible that where ivory is present alongside skulls, elephants may, through tactile or olfactory cues, recognize tusks from individuals that they have been familiar with in life.

The evolutionary basis for exhibiting such intense interest in the decomposed remains of con-specifics in a nonhuman mammal remains unclear. While the behaviors described here obviously differ fundamentally from the attention and ritual that surround death in humans, they are unusual and noteworthy. Comparative research is now required to test systematically whether any other species show similar responses and what relationship, if any, they have to particular cognitive abilities or aspects of social behavior.

**EMPATHETIC
AND ENDANGERED
ELEPHANTS**

The Good Samaritans

In June 2000 Cornell University bioacoustician Katy Payne and other scientists were sitting perched on a platform above a forest clearing in the Dzanga-Sangha Forest Reserve of the Central African Republic watching elephants. The animals had emerged from the surrounding forest for their regular visit to the mineral-rich mud wallows of the clearing.

One day, a young elephant, weak from malnutrition, collapsed off to one side of the narrow, sandy trail leading to the clearing. She lost consciousness and within a few hours died, and her collapse and death were witnessed by her mother and sister. The scientists on the platform also observed the event. And they documented, using a video camera with a telescopic lens, the reactions of a large number of elephants as they ambled along the trail to and from the clearing. In that fashion, the scientists acquired an extensive record of elephant behavior in response to a dying fellow elephant (on day 1) and a dead fellow elephant (on day 2). In total, the elephants paid 129 visits to the fallen animal.

Most of those visits began as you might expect: with investigation, exploration, sniffing the body, gently touching it, and so on. What next? About 50 percent of the visits ended with a response of fear or avoidance: backing off, sidling away, or dashing off. That's not surprising, since the presence of a stricken elephant in a place frequented by elephant poachers can indicate danger. One elephant, a notorious misfit, reacted aggressively by stabbing the body with her tusks. But on a third of the visits the elephants showed neither fear nor aggression but helping behavior. In roughly half of those cases, an elephant tried to protect the dead or dying animal from others. In the other half of helping-behavior cases, the visitor actually tried to rescue the fallen elephant—for example, using feet, trunk, and tusks in an effort to raise her back up on her feet.

Other researchers (particularly Andrea Turkalo, the founder of the Dzanga-Sangha elephant project) had already watched these elephants for years, identifying individuals and charting familial and social relationships. Examining this earlier data, Payne and her colleagues found no clear correlation between how individuals reacted and their relationship to the unfortunate elephant. Strangers were as likely to try defending her, or rescuing her, as relatives or close acquaintances were. This, then, seemed to be a case where none of the standard evolutionary logic of cooperation applies. It was more a case of spontaneous kindness or altruism: elephants trying to defend or assist a fellow creature in trouble without nepotistic bias or the anticipation of a mutual or reciprocal benefit.

It also provides an intriguing animal counterpart to the biblical parable of the Good Samaritan. To refresh your memory: A man traveling on the road between Jerusalem and Jericho is attacked by robbers, who take all he has, strip him of his clothes, and leave him to die alongside the road. As the man lies there, helpless, bleeding, naked, dying, three other travelers discover him. Two are seemingly respectable men, well-established members of the same social and religious group Jesus was addressing when he told the story. These two men notice the dying man, but—perhaps appropriately fearful of thieves in this lonely place—they cross to the other side of the road and hurry along. That is a reasonable response, one that you and I can identify with. It's dangerous there.

All the more remarkable, then, was the response of the third traveler, who happened to be a Samaritan, a member of that despised and distrusted set of apostate Jews who had developed their own version of the Torah. What would a Samaritan know about goodness or kindness or personal ethics? Yet when this man saw the robbers' victim, he stopped and attended to him. He washed the man's wounds, wrapped them with bandages, and then placed the man onto the back of his own pack animal and transported him to a nearby inn. The Good Samaritan paid for the victim's lodging and, on leaving the next day, gave the innkeeper two silver coins to pay for the victim's care, then promised to cover any additional charges when he returned.

As Jesus indicated, this didactic tale provokes us to consider our duty to our neighbor. Our duty is to behave not like the two respectable men who, listening to their own fears, reacted to the grievously injured man with avoidance. No, our duty is to emulate the Samaritan, who embodied a central tenet of Christian teaching, which is to practice radical kindness—to love one's neighbor as oneself.

That's the Good Samarian story: a narrative of ethical choice that can help us think about the differing behaviors exhibited by those elephants who, walking along the Dzanga-Sangha trail one summer day, discovered a fellow elephant in trouble. Most of them did the sensible thing, which was to respond cautiously and then move away. Some of them did the less sensible thing, which was to ignore or suppress their fear and respond with compassion.

The idea that elephants can experience compassion will probably not surprise most elephant experts, nor will it shock many experts in animal cognition and behavior, although they might substitute the slightly more specialized word "empathy" for my more common and generic term "compassion." They would also recognize that the emotion or emotional complex of empathy can underlie the behavior sometimes known as *altruism*.

We too often imagine that humans, with their marvelous brains and powerful cultures, can construct their own worlds and manufacture their own moral codes—and thus rise above the tooth-and-nail, survival-of-the-fittest world of nonhuman animals. How could evolution possibly have produced the emotion of empathy and the behavior of pure altruism? Altruism, after all, seems to be a loser's strategy. It's true that altruism is admirable, inspirational. We honor the woman who sacrificed her life so that another human being, a perfect stranger, might live. But the woman who sacrificed her life has also sacrificed her genetic future. Any genes she possessed have not moved into the next generation. Over time, wouldn't the infinitely patient winnowing of evolution simply cast away all genes associated with self-sacrifice? How does the competition of evolution even begin to explain a dangerous act done for a nonrelative?

One possible answer to this conundrum is to imagine the woman's self-sacrificial altruism as an extraordinary negative consequence of a set of genes that is ordinarily positively useful. It could be that for every person who has died a hero, another dozen equally passionately altruistic individuals have survived and found their social and economic status elevated as a result of their heroism.

A more intriguing answer has recently been proposed by primatologist Frans de Waal and cognitive neuroscientist Stephanie Preston. De Waal and Preston promote a model that avoids treating altruism and empathy as isolated phenomena. Instead, their model would combine helping behavior with empathy, perceptual and emotional contagion (the spread of yawning in a classroom, for example), and a few similar phenomena into a unified whole. Each element of this whole is, or could be, the product of a single empathetic neural system. The recent discovery of specialized brain cells known as mirror neurons—cells that seem to automate an individual's perceptual and emotional response to many kinds of social stimuli—strengthens the case for a dedicated empathetic neural system that might be distributed among many species.

Once we see empathy and altruism as part of a larger whole, we are also prepared to recognize the benefits provided by the full system. For example, this system might make mothers instantly and automatically—contagiously—responsive to the needs and actions of their infants. It may allow many members of a social group to benefit from the heightened alertness of the group as a whole: with contagious processes transforming ten individual animals, each with only two eyes, into a coherent living entity with twenty eyes. It would help coordinate group actions quickly and automatically, thereby improving any individual's chances of survival from predators. The following excerpt is taken from Payne's scientific account of the Good Samaritan elephants of Dzanga-Sangha.

⎯ ⎯ ⎯ ⎯ ⎯ ⎯

The following events took place in the Dzanga Sangha forest clearing in the Central African Republic, where Andrea Turkalo has been documenting the demography and behaviors of a population of African forest elephants for a decade. Several colleagues and I joined Turkalo for three months in spring 2000. On June 26, we observed and videotaped the death of a yearling elephant calf on a spot about one hundred meters from our observation platform, and (on that and the next day) 129 visits to the body by elephants of both sexes and all ages.

The cause of death was probably starvation. We had observed the calf with her mother (named "Morna I" in Turkalo's demographic study) and a well-fed older sister several times in the weeks preceding death, noting that the mother was rejecting the infant's attempts to nurse while favoring the sister's attempts. As her calf died, within two meters of a much-traveled elephant trail, the mother and sister stood over or near her. Both tended to draw aside when other elephants approached the trio on the path. The visiting elephants tended to arrive one at a time on a hard sandy path flanked by soft mud; thus their perceptions and responses were to an unusual extent independent of one another.

The last visible movements of the calf occurred about two hours before the end of the first observation period. Not knowing exactly how to determine the moment of death, we have divided the data into the "day of death" and "day after death."

On the day of death we documented 56 visits to the body by 38 individuals, including the calf's mother and sister, who made 6 of the visits. On the next day we documented 73 visits by 54 individuals. The calf's mother and sister had presumably left the area: we did not see them again during our remaining 22 days of observation.

Of the 129 visitors, 128 changed their behaviors as they approached the body. The behavioral changes included: exploratory behaviors in 80 percent of the encounters (obvious air and body sniffing, hastened approaches to body, touching or hovering foot over body, touching body with trunk, tasting trunk after body touch); fear/alarm behaviors in 50 percent of the encounters (sidling or faltering approaches, backing away from body altogether, avoidance in detours off all paths, ears lifted, tails lifted, trumpeting and rumbling, body jolts, hasty departures after exploration); efforts to lift the dying calf in 18 percent of the encounters, using foot, trunk, or tusks; body-guarding reactions in 15 percent of the encounters (threatening others—including the calf's mother—away from the body and standing over or next to it for a period of time); and aggression toward the body (tusk-stabbing and ripping the body apart) in a few prolonged visits by one individual, a subadult female who had in the previous year mauled a dying calf, and whose behaviors toward human observers were also aberrant.

The nature, extent, and coupling of these responses differed dramatically from individual to individual, as did the time spent attending the body and the number of returns to the body. The most extreme responses were given by a subadult male, Sappho II, and a young nulliparous adult female, Miss Lonelyheart. Sappho II visited the body five times (a total of thirteen minutes), specializing in exploration and potential succorance, including fifty-seven attempts to raise the calf to her feet. Miss Lonelyheart (so named because some of her other behaviors had been noted as bizarre since 1998 when she was first identified) spent 250 minutes at the body, tearing it apart and occasionally putting bits of it into her mouth. This individual was the only one showing aggression toward the calf. Other visitors expressed a broad range of fearful, affiliative, and exploratory behaviors, in various combinations. Morna I, the calf's mother, was among the individuals who spent the most time with the body and performed the most affiliative behaviors, exceeded in the first capacity only by Miss Lonelyheart and in the second only by Sappho II.

Long-term records indicate dates on which identified individuals have been seen in the Dzanga clearing. An association matrix was made from these, and used to evaluate the relative probability of genetic or social relationship between the dead calf's mother and each of the elephants identified in the clearing on June 25–26. A comparison of association levels of those who did and did not visit the body yielded a P-value of .544 (two sample t-tests), indicating that visiting elephants were not more closely associated with Morna I than the elephants in the clearing who did not visit the body. This confirms our impression that first encounters with the body were, except in the case of the mother and sister, the consequence of the body's location next to a well-used trail. Returns to the body were, of course, another matter, suggesting interest and intention.

Of all visitors, only Morna's two calves, who were always with her, had coincided with her in more than 27 percent of their appearances at the clearing throughout the period since Morna's first identification in 1991. In spite of these low levels of association, one might expect some correlation between the nature and intensity of visitors' responses and their history of association with Morna I. One might also expect males to be less responsive than females.... However, [we documented] an absence of relationship between Morna I–visitor association levels and each of three factors: the time spent at the body, the number of returns to the body, and the number of affiliative behaviors performed. Gender did not offer a predictor of behavior either. This series of encounters seems, then, to offer evidence of individual variability beyond what one would expect as the consequence of differences in age, sex, and degrees of relatedness. Probably it is evidence of variability in several interactive domains, including cognition, emotion, and the physiological factors that account for levels of reactiveness.

The value of this example lies in its documentation in elephants of what we would broadly refer to, in human subjects, as personality differences. Undoubtedly, such differences are also reflected in elephants' more ordinary behaviors, but rarely does one get the chance to eliminate the contribution of such factors as age, sex, and relatedness as cleanly as in this example. We cannot, however, assess the extent to which these idiosyncrasies may have been shaped by prior experience. Given their long lives, their social memory, and their capacity for learning, it would be remarkable if elephants did not develop conspicuous personality differences in the ordinary course of their lifetimes.

The responses illustrated in this example take at least one form that seems maladaptive, namely Miss Lonelyheart's persistent attacks on an unrelated female's dying calf. This unexpected dimension sheds new light on the importance of individual recognition and remembered interactions—adaptive traits for members of a society in which sociopathic behaviors can develop. Here we see another way in which the increasing social memory of a matriarch probably contributes to the inclusive fitness of her family.

Rescuing the Antelopes

LAWRENCE ANTHONY

In *The Elephant Whisperer* (2009), Lawrence Anthony tells his story of bringing a troubled herd of elephants into the Thula Thula wildlife reserve of Zululand, a province of South Africa. This herd of nine elephants had been taken from another wildlife reserve six hundred miles to the north because they had proven themselves to be skilled escape artists, capable of breaking through high-voltage electrical fences. In the process of their removal from that earlier reserve, the herd's matriarch and her infant son were shot and killed, and when the enormous van carrying a seven-elephant herd finally arrived at Thula Thula and the animals were released into a very large *boma*, or holding pen, they were severely traumatized and raging with fear and hostility.

Within a short while after their arrival, the elephants had pushed over a tree to short-circuit the high-voltage wires surrounding the boma. Then they plowed their way through the concrete- and cable-reinforced fencing and were running loose in the five-thousand-acre reserve. They followed the miles of electrical fencing surrounding the reserve until they had traced the current back to its source, a remote charger hidden in a woodsy thicket a half-mile away from the fence. They demolished the charger. With the electricity down, they returned to the eight-foot-high outer fencing and pushed it over, then headed north in the direction of their previous home. Using a helicopter, experts from the Zululand Wildlife Service located the elephants, knocked them unconscious with anesthetic darts fired by a marksman in the helicopter, craned them one by one into a transport van, and returned them to the boma at Thula Thula for one last chance. Next time, the authorities declared, the entire herd would be shot as a public menace.

Anthony, as owner and manager of Thula Thula, decided that the only way to save these elephants was to develop a relationship with their leader, the new matriarch. He set up a permanent camp next to the boma, staying there day and night, communicating his presence to the elephants by talking, singing, whistling, or simply keeping himself visible outside the fence. The high-voltage wiring of the boma can knock a big man flat on his back, and it would cause serious pain to any elephant who touched it. But an elephant who knew enough to tolerate the initial pain for a few seconds could rip out the wiring and break the circuitry. It seemed clear to Anthony that the new matriarch of this herd—named Nana—was prepared to do exactly that. Every morning at 4:45 a.m., in the darkness just before dawn, Nana and the rest of the herd would silently gather at the northern end of the boma, massing together and preparing, it seemed, to smash through the eight-foot-high fence. Every morning at the same time, Anthony stood on the outside of the fence and spoke to the giant matriarch looming above him, pleading with her not to do it.

He never expected her to understand the words, but he believed she might understand the feelings behind the words. Indeed, after several minutes of this sort of drama each morning, she would turn around and disappear into the darkness, followed by the rest of the herd. Within a few weeks it seemed that the matriarch's hostility had dissipated. One morning she and her baby quietly approached the part of the fence nearest Anthony's camp, and suddenly, as he later wrote, "I felt sheathed in a sense of contentment. Despite standing just a pace from this previously foul-tempered wild animal who until now would have liked nothing better than to kill me, I had never felt safer." She reached out with her trunk, and he understood that she wanted him to approach the fence. "Time was motionless as Nana's trunk snaked through the fence, carefully avoiding the electric strands, and reached my body. She gently touched me....After a few moments I lifted my hand and felt the top of her colossal trunk, briefly touching the bristly hair fibres."

The message Anthony understood was that Nana and the rest of the herd were ready to be released from their prison. He turned off the electricity and opened the gates. Nana walked out, followed by the others. The herd thus became seven free-ranging members of the wildlife community at Thula Thula, which included hippos, rhinoceroses, giraffes, zebras, hyenas, warthogs, the occasional lion, and numerous antelopes small and large. Among the largest and most dramatic of the antelopes were the nyala: spiral-horned ungulates indigenous to southern Africa—the males slate gray, the females and young rusty brown with white stripes—who can weigh up to three hundred pounds.

A few years after the elephants arrived, Anthony, having noted an abundance of nyala at Thula Thula, decided to capture thirty of them and transport them to another wildlife reserve for breeding. He had a small boma constructed for the antelopes, hired a specialist to capture the animals, and sent out his manager, Brendan Jones, and several workers to run the capture operation and watch over the nyala at night. The following passage from *The Elephant Whisperer* describes what happened next.

T he numerous nyala grazing literally outside our bedrooms re-minded me that we had a surplus of these magnificent antelope on the reserve and I decided we should sell about thirty off to other reserves for breeding purposes.

A phone call later and a game-capture specialist was on Thula Thula darting the animals, which we placed in a *boma* with plenty of fresh water and alfalfa until we had reached the sell-off quota. We would then load them into the customized van and he would deliver them to the buyer.

Brendan was overseeing the capture and radioed to say we had our quota and the van would be leaving the next morning. It had been a long day. I was tired and looking forward to an early night. Thus I was surprised to be woken by a radio call from Brendan at 11 p.m. "You'd better come down. The most amazing thing has just happened."

I cursed, pulled on some clothes and drove down to where Brendan and the team were waiting. The first thing I noticed was that the door to the *boma* was open.

"Where're the nyala? Surely you didn't load at night!"

I turned to the game-capture man who was standing with his staff staring at the open door. He looked as though he had seen a ghost.

"You're not going to believe what happened," he said.

"Try me!" My patience was somewhat aggravated by lack of sleep.

"We were sitting by the *boma*, just chatting," he said, "when we heard the elephants come. A couple of minutes later Nana led the herd into the clearing and so we moved right off—some quicker than others," he grinned, looking at Brendan. "We thought she had smelled the alfalfa. We had twelve buck inside and we were worried what would happen to them if the *boma* was flattened by the herd going crazy for the food.

"Then the herd stopped, as if on instruction. Nana walked alone to the *boma*. Just as we thought she would smash through the fence, she stopped at the gate. It wasn't locked because the clasps were folded and were secure enough. She started fiddling with the clasps and got one open, then the other, and then pulled open the door. We couldn't believe it, she actually opened the damn door!"

He looked around as the others nodded.

"Then instead of going for the alfalfa, which we thought was her whole mission, she stood back and waited. After a few seconds a nyala came out, then another, and before we knew it they had all found the gap and were gone.

"The weirdest thing is that as the last one fled, Nana just walked off and the others followed. They didn't even go for the alfalfa—a pile of prime chow and they just ignored it."

I looked at him, smiling. "Okaaaay. So what you're saying is that the elephants felt sorry for the poor old nyala. They came across the reserve just to release them out of the goodness of their hearts. Because they had nothing better to do. Good try. Now...what really happened?"

"I swear to God that's exactly what happened. Ask the others." And with that they all started jabbering away simultaneously, backing up him and outdoing one another in verifying the story.

It took me a bit of time to digest it but there was no doubt they were telling the truth. There were elephant tracks all around the *boma* and Nana had thoughtfully dumped a steaming pile at the gate as a smoking gun. The lock clasps were also all smothered in trunk slime.

How or why this occurred remains a mystery for some, but it's a mystery only if you grant elephants limited intelligence. Once you grasp that these ancient giants who have roamed the planet since time immemorial are sentient beings it all becomes clear. Nana, once a prisoner of the *boma* herself, had decided to let the nyala go free. It is as simple—or complicated, if you like—as that. There can be no other explanation.

Scents and Sensibilities

LYALL WATSON

The dual subject of Lyall Watson's *Elephantoms: Tracking the Elephant* is empathy and trauma, the kind of empathy elephants are capable of and the sort of trauma that can make them vulnerable. The book begins, though, with Watson's recollections of growing up during the 1940s in South Africa, an era when it was possible to spend four weeks each summer living with a collection of boys his age—a minor tribe of sorts, no adults allowed—in a driftwood hut on a beach facing the Indian Ocean at the southern tip of the continent. The boys survived as subsistence foragers, more or less, and they called themselves Strandlopers, recalling the term for Beachwalker once applied to beach-living hunter-gatherers sighted by the first Dutch settlers to arrive at the Cape in 1652.

Obviously, these Strandlopers were modern and privileged versions of the original concept. Descendants of Europeans, they were on vacation and within walking distance of a telephone in case of emergency, provisioned with a cache of matches, candles, fishing gear, soap, flour, and sugar. They were also blessed with plenty of fresh fish—and yet constrained by an absence of fresh water, since it seldom rained during the summer weeks. Sometimes, when a morning mist rolled in heavily, they would wash themselves and drink by sucking on wet grass and leaves. Sometimes they discovered water by following the trails of bush pigs, who were especially adept at sniffing it out. Other times, though, the shortage of water forced them to range farther afield, and

241

it was during one of their extended water-seeking expeditions that Watson and two fellow Strandlopers were struck by an extraordinary vision: a massive male elephant with enormous curved tusks standing where elephants were not supposed to exist, a great creature colored in a way elephants were not expected to be colored—that is, strangely pale, "like patinated bronze or the color of old ivory." The animal appeared mysteriously, then turned silently and vanished casually in an instant. He seemed, in vanishing, more fantasy than reality, an elephant phantom, or "elephantom."

At the time of that first elephantom, perhaps a dozen elephants remained in the entire ecosystem of Cape forests that, draped like a ragged cloak around the southern logging town of Knysna, had once covered three thousand square miles and harbored hundreds or thousands of wild elephants. Indeed, of the estimated 100,000 elephants roaming free in southern Africa when the first Europeans arrived in the seventeenth century, little more than a hundred free-ranging individuals remained by 1920. Most of the others had been slaughtered by white men with guns. In his book, Watson imagines a continent filled with elephantoms, a haunting presence that lingers in the way invisible clouds of scent molecules might linger after a killing or crime, for days or weeks or years.

As a young adolescent, Watson searched for wild elephants elsewhere in southern Africa. He went on to acquire degrees in botany and paleontology from the University of Witwatersrand before moving to Europe to study anthropology in Germany and Holland, then to work with the famed animal behaviorist Desmond Morris and complete a doctorate in ethology at the University of London before returning at last to South Africa as the director of the Johannesburg Zoo. In *Elephantoms*, Watson presents a lyrical, flowing memoir interwoven with astonishing explorations on the subject of elephants vanished and vanishing—of what we know about them, historically and scientifically, and what we don't know. In the following excerpt from the book, he describes his experiences at the Johannesburg Zoo with an elephant named Delilah, an orphan, possibly the traumatized survivor of a cull, who despite her early trauma was "one of the sunniest, most consistently good-tempered individuals" he had ever known.

——— —— —— —— —— —— —— ——

The Johannesburg Zoo was everything London Zoo was not. It was a municipal institution, short of cash, long on bureaucracy, and run by the Parks and Recreation Department. Which meant that the lawns were manicured, the flower beds lovely, but the animal quarters disastrous. I was the first professional zoologist to be employed there. Most of the staff were untrained, largely uninterested, and entirely white, working there only because no other department would have them.

It was an uphill struggle. Far too many pointless meetings, too much talk, and everything else in triplicate. Requisitions were a nightmare; it was always so much easier to say no rather than have to come and see problems for themselves "out at the zoo." We always seemed to be last on everyone's list of priorities. And then the only one who really cared, the man who had found me and hired me in London, left his post and we lost whatever clout we had, along with most of our budget. But then there was still Delilah...

She was four years younger than I, a teenager, born in the bush, but having lived most of her life in Johannesburg. She was an orphan, the survivor of a massacre, but despite this background she was one of the sunniest, most consistently good-tempered individuals I have ever met. She was also truly beautiful, with long, thick eyelashes. And I was particularly fond of her trunk.

Delilah lived alone in a dark, damp, concrete-floored cage in what was euphemistically described as the Elephant House. More than half of her time was spent shackled indoors, chained to a ring in the floor that gave her the scope of just eight feet of chain. During the day, she had the "free run" of a compound half the size of a tennis court surrounded by girders of black steel bent out of shape by earlier, angrier denizens.

This was where I first met her, standing near her steel barrier, rocking gently in a way I had learned to recognize as one of the first signs of stress and mental illness. As I approached, she pushed out her trunk directly toward me in the gesture all elephants use on meeting strange or higher-ranking individuals. I knew, thanks to Desmond [Morris], that this was a "greeting-intention movement," one universally misinterpreted in zoos as "begging" and rewarded by offers of food, when what was really being sought was friendship. So I cupped the tip of her trunk in my hand and gently blew into it.

The result was extraordinary. She entwined my whole arm in her trunk, held it there as she breathed deeply several times, and then put the tip of her trunk in her mouth and sighed. I came a little closer and let her explore my face and neck freely until I could hear a soft growl of pure delight—the elephant equivalent of purring.

It was love at first sight, and I decided, then and there, that my first priority in this zoo would be a new elephant house and company, elephant company, for Delilah.

That took time, but as construction continued, I got the chance to get to know Delilah a great deal better.

To start with, her initial automatic, head-lowered, ears-flattened, swaying gesture of submission made me curse the early keepers who must have beaten her into obedience. But as we became better acquainted I was happy to see more confident approaches, raising her head, tucking her chin in, and lifting and flapping her ears—all in a short gallop to the fence, even before we had gone through the snaky business with our trunks.

I never fed her. That is not something elephants normally do for one another. I left that to her regular keeper, who by now was beginning to take a closer interest in her welfare. And eventually I decided it was time to go into the compound with her, alone. Just me and over eight thousand pounds of female elephant on a blind date.

To get into the outdoor area, I had to go through the store and the indoor area—the usual way keepers approached her, with all the usual sounds. But by the time I stepped out of the darkness into the sun, she was already taking unusual interest. She had heard, smelled, seen something other than her keeper and was *standing tall*. She flapped her ears and lifted her head very high, trunk spelling out the letter S in front of her head, the tip swiveling my way in full alert, a thin dribble of dark fluid on each cheek. "Oh-oh!" I thought. "This could go wrong. Have I miscalculated?"

She began to move toward me somewhat stiff-legged, trunk now hanging at a more acute angle, but still not showing the side-to-side head shake of outright threat. That would have sent me back indoors in a hurry. Then I heard the door close behind me, cutting off any retreat. The keeper obviously didn't want to be involved in any of this, or he wasn't very fond of me. Still she kept coming, keeping me guessing until the last moment, when she stopped right in front of me and very deliberately pressed the top of her trunk against my forehead so that I could feel a soft vibrant rumbling sound right through my body. She was leaning into me, purring something that sounded very much like "Hey! What took you so long?"

For a while, construction stopped altogether—something about the supply of cement—and I wondered what else I could do to keep Delilah amused. I contemplated bringing back elephant rides. There were still a pair of brick ladders up near the War Memorial, gangways like those on airport aprons, where children once boarded a howdah—a sort of saddle with seats—to be taken once around the swan lake for sixpence. It hadn't been used since an animal welfare organization made a fuss, condemning it as demeaning to elephants. They were right, it is, but for elephants stuck in a cage it could have been a welcome change of pace. I was sure Delilah could be taught to wear the howdah and would enjoy the company, but the city fathers and their lawyers squashed the whole idea—and I got cement in a hurry.

In the meantime, and before zoo opening time, I took to walking Delilah around myself. We used a form of bridle with a leather lead which both she and I pretended would keep her in line. It was never tested, for the simple reason that she really enjoyed walking around with me. She wasn't fond of monkeys or little creepy things like honey badgers and porcupines, so we avoided that part of the zoo and strolled instead between the paddocks of zebras, wildebeests, and giraffes. These seemed familiar to her and she spent a lot of time with her trunk hanging over their fence, trying to remember where and when they had met. I'm not sure she ever did make the connection. She was only three years old when she was captured, and it can't have been more than just a faint memory of happy times with the herd.

Sometimes she let me grip her tail while she decided which way to go, trotting along, squeaking like a calf, reliving perhaps those times when her mother steered her with a firm trunkhold on her behind. And I could have sworn she found pleasure in this strange reversal of roles. But there was another day when this game nearly backfired.

Delilah must have heard lions before. There were several in the zoo and they roared almost every evening, giving people who lived in the crowded suburbs nearby a *frisson*, reminding them that, all immediate appearances to the contrary, they were still living in Africa. I am certain that lion smell had been part of her zoo experience for sixteen years and wouldn't normally disturb her. But I had forgotten, or was never told, that a new male lion—one said to look very like the extinct Cape lion—was being brought in that day and would be taken to the Lion Hill in our transport cage on wheels.

I even heard the tractor trailer coming our way, but this was such a normal part of zoo routine that I never thought twice about it until it was just twenty feet away and the lion flung himself at the bars with a deafening roar. That sound, anywhere nearby, is enough to turn your knees to water. In Delilah, it triggered an instinctive response. She whirled around and put herself between me and the lion, doing everything possible to assume a group defense all on her own. There were no signs of indecision, no trunk coiling or winding, no ear-touching or pulling up tufts of grass. Her tail stiffened in my hand, her back arched, her head shot up, and her ears spread out to their full extent, providing an awesome frontage of ten or twelve feet of gray anger studded with tusks and accompanied by an ear-splitting scream from her raised trunk.

Even from behind it was impressive. From the lion's side, it must have been absolutely terrifying. It took us two days to get him out of the traveling cage and an entire week before he dared to show his nose in the outdoor enclosure!

Delilah took it in her giant stride. For weeks afterward, whenever she thought no one was looking, I saw her replay her display, polishing some of the moves. And at the end of each silent rehearsal, she adopted that funny, loose-limbed sort of swagger that in the elephant world invariably indicates a large degree of self-satisfaction.

When the domed buildings of the new elephant area were almost complete, we got word of a pair of young elephants who had survived a cull on the border with Botswana. I pulled a few strings, and within the week they were ours and on their way to Johannesburg.

The plan was to keep them in the old building next to Delilah so that they could get used to each other through the bars before they were all turned loose in their new home. Delilah's part of the original building was larger, better equipped for the two newcomers, so we moved her to the smaller older wing, which had not been used at all since her arrival.

She was reluctant to make the move and had to be led by hand into the wing, moving very slowly, step by step, hanging back as long as she could. The place had been spring-cleaned, scrubbed and furnished with fresh hay and water, but it was clear that she didn't like it. I stayed with her all the way, making encouraging noises, but that didn't help much. Maybe it was my accent. In the end, however, she settled down a little and we left her to it as I watched from behind the scenes.

She started sniffing first at the food and bedding, and then moved across to the other side of the indoor area, the tip of her trunk opening and closing, testing smells left and right, reaching out to its full extent as she got closer to the wall. Then the pattern changed, she began to concentrate on one spot in the corner, pausing, turning, hesitating, finally giving all her attention to that small area. She became very quiet, even tense, and stood right over the spot, giving it her undivided attention, so absorbed that even her trunk stopped moving. And she stayed that way, entranced, for minutes on end.

Everything about her demeanor reminded me of the young bull I had seen in the Addo investigating another elephant's skull. Eventually, Delilah shook herself out of the meditation and seemed to come to a decision. She went over to the hay pile, picked up a large sheaf with her trunk, and carried this across to the offending area. And she kept on transporting hay until the entire corner was completely concealed. Then she relaxed and seemed quite at home.

I called the keeper and showed him what she had done. It didn't make any sense to him either until I asked how long it was since the wing had been used.

"*Gits,*" he said. "Almost twenty years. This is where we kept the last African elephant. The one we had before Delilah arrived...."

I asked what had happened to it.

"She became very sick and difficult and had to be kept shackled all the time. Until eventually the visiting vet said she would have to be put down. We shot her..."

He paused and I could see that something had just occurred to him.

"*My God,*" he said, with his eyes wide. "That's where it happened, all those years ago. That's where she died. Right in that corner!"

Blood Ivory

BRYAN CHRISTY

Fifteen years ago, while driving across a stretch of sweet-smelling grasslands at the edge of forest in Gabon, Central Africa, I saw in the distance a great splash of white that looked like a riotous tumble of white linen. On closer examination, the bright splash became a dull scattering of giant bones: the discarded trash of an ivory crime. Within that scattering, I found a jawbone so tiny and yet so perfectly formed that it must have come not from an elephant baby but rather from an elephant fetus. I had been stumbling around the remains of a pregnant female, I concluded, who had been about to give birth but was instead slaughtered on the spot by some ferocious human being.

People have killed elephants for their ivory for many thousands of years, but only today, only in our historical moment, is it possible to imagine we are reaching the end of it. We are reaching the end of the killing because we are reaching the end of elephants. The most thorough scientific study of African forest elephant trends, published in 2013, indicates that the total population of African forest elephants declined by nearly two-thirds in a single decade, from 2002 to 2011, as a result of ivory poaching. (African forest elephants are distinctive enough from African savanna elephants to be considered, by many experts, as a separate species.) A second scientific study, released in 2014 and arguably the most thorough quantitative study so far, suggests that some 100,000 African elephants of both species were killed by poachers in just three years. In a single year, 2011, poachers had slaughtered one out of every twelve elephants alive in Africa.[48]

This is a cataclysmic slaughter, a massive crime against elephants, and because it threatens the actual existence of a taxon of intelligent and iconic animals who continue to inspire and illuminate us and our children and grandchildren, it is also a great crime against humanity. The criminals are the poachers, of course, but the criminals are also the big men who supply and protect the poachers; the transporters and the dealers and the smugglers who move the ivory; the nations and organizations enabling that movement; and finally, the ivory consumers themselves—criminals who somehow are unable to smell the stink of blood on the precious objects they have so eagerly acquired.

One of the best contemporary reports on blood ivory, Bryan Christy's "Ivory Worship," which appeared in *National Geographic* magazine, provides a wide overview of the ivory business that reminds us not only of the original criminals and the ongoing slaughter in Africa they are perpetrating at the supply end of the business, but also of the supporting criminals who buy and sell it at the demand side. Those supporting criminals include Catholic priests in the Philippines and Buddhist monks in Thailand. The following excerpts from Christy's larger report describe a Roman Catholic hierarchy in the Philippines who link their extraordinary trade in devotional objects carved from ivory with Muslim smugglers who bring the precious raw material out of Africa. Like the Philippine hierarchy, Christy writes, the Vatican also continues to support the buying and selling of blood ivory for profit and the glory of God.

In January 2012 a hundred raiders on horseback charged out of Chad into Cameroon's Bouba Ndjida National Park, slaughtering hundreds of elephants—entire families—in one of the worst concentrated killings since a global ivory trade ban was adopted in 1989. Carrying AK-47s and rocket-propelled grenades, they dispatched the elephants with a military precision reminiscent of a 2006 butchering outside Chad's Zakouma National Park. And then some stopped to pray to Allah. Seen from the ground, each of the bloated elephant carcasses is a monument to human greed. Elephant poaching levels are currently at their worst in a decade, and seizures of illegal ivory are at their highest level in years. From the air too the scattered bodies present a senseless crime scene—you can see which animals fled, which mothers tried to protect their young, how one terrified herd of fifty went down together, the latest of the tens of thousands of elephants killed across Africa each year. Seen from higher still, from the vantage of history, this killing field is not new at all. It is timeless, and it is now.

The Philippines Connection

In an overfilled church Monsignor Cristobal Garcia, one of the best-known ivory collectors in the Philippines, leads an unusual rite honoring the nation's most important religious icon, the Santo Niño de Cebu (Holy Child of Cebu). The ceremony, which he conducts annually on Cebu, is called the Hubo, from a Cebuano word meaning "to undress." Several altar boys work together to disrobe a small wooden statue of Christ dressed as a king, a replica of an icon devotees believe Ferdinand Magellan brought to the island in 1521. They remove its small crown, red cape, and tiny boots, and strip off its surprisingly layered underwear. Then the monsignor takes the icon, while altar boys conceal it with a little white towel, and dunks it in several barrels of water, creating his church's holy water for the year, to be sold outside.

Garcia is a fleshy man with a lazy left eye and bad knees. In the mid-1980s, according to a 2005 report in the *Dallas Morning News* and a related lawsuit, Garcia, while serving as a priest at St. Dominic's of Los Angeles, California, sexually abused an altar boy in his early teens and was dismissed. Back in the Philippines, he was promoted to monsignor and made chairman of Cebu's Archdiocesan Commission on Worship. That made him head of protocol for the country's largest Roman Catholic archdiocese, a flock of nearly 4 million people in a country of 75 million Roman Catholics, the world's third-largest Catholic population. Garcia is known beyond Cebu. Pope John Paul II blessed his Santo Niño during Garcia's visit to the pope's summer residence, Castel Gandolfo, in 1990.

Some Filipinos believe the Santo Niño de Cebu is Christ himself. Sixteenth-century Spaniards declared the icon to be miraculous and used it to convert the nation, making this single wooden statue, housed today behind bulletproof glass in Cebu's Basilica Minore del Santo Niño, the root from which all Filipino Catholicism has grown. Earlier this year a local priest was asked to resign after allegedly advising his parishioners that the Santo Niño and images of the Virgin Mary and other saints were merely statues made of wood and cement.

"If you are not devoted to the Santo Niño, you are not a true Filipino," says Father Vicente Lin Jr. (Father Jay), director of the Diocesan Museum of Malolos. "Every Filipino has a Santo Niño, even those living under the bridge."

Each January some 2 million faithful converge on Cebu to walk for hours in procession with the Santo Niño de Cebu. Most carry miniature Santo Niño icons made of fiberglass or wood. Many believe that what you invest in devotion to your own icon determines what blessings you will receive in return. For some, then, a fiberglass or wooden icon is not enough. For them, the material of choice is elephant ivory.

Filipinos generally display two types of ivory *santos:* either solid carvings or images whose heads and hands, sometimes life-size, are ivory, while the body is wood, providing a base for lavish capes and vestments. Garcia is the leader of a group of prominent Santo Niño collectors who display their icons during the Feast of the Santo Niño in some of Cebu's best shopping malls and hotels. When they met to discuss formally incorporating their club, an attorney member cried out to the group, "You can pay me in ivory!"

I tell Garcia I want to buy an ivory Santo Niño in a sleeping position. "Like this," I say, touching a finger to my lower lip. Garcia puts a finger to his lip too. *"Dormido* style," he says approvingly.

My goal in meeting Garcia is to understand his country's ivory trade and possibly get a lead on who was behind 5.4 tons of illegal ivory seized by customs agents in Manila in 2009, 7.7 tons seized there in 2005, and 6.1 tons bound for the Philippines seized by Taiwan in 2006. Assuming an average of 22 pounds of ivory per elephant, these seizures represent about 1,745 elephants. According to the Convention on International Trade in Endangered Species of Wild Fauna and Flora (CITES), the treaty organization that sets international wildlife trade policy, the Philippines is merely a transit country for ivory headed to China. But CITES has limited resources. Until last year it employed just one enforcement officer to police more than thirty thousand animal and plant species. Its assessment of the Philippines doesn't square with what Jose Yuchongco, chief of the Philippine customs police, told a Manila newspaper not long after making a major seizure in 2009: "The Philippines is a favorite destination of these smuggled elephant tusks, maybe because Filipino Catholics are fond of images of saints that are made of ivory." On Cebu the link between ivory and the church is so strong that the word for ivory, *garing,* has a second meaning: "religious statue."

The Catholic-Muslim Underground

"Ivory, ivory, ivory," says the saleswoman at the Savelli Gallery on St. Peter's Square in Vatican City. "You didn't expect so much. I can see it in your face." The Vatican has recently demonstrated a commitment to confronting transnational criminal problems, signing agreements on drug trafficking, terrorism, and organized crime. But it has not signed the CITES treaty and so is not subject to the ivory ban. If I buy an ivory crucifix, the saleswoman says, the shop will have it blessed by a Vatican priest and shipped to me.

Although the world has found substitutes for every one of ivory's practical uses—billiard balls, piano keys, brush handles—its religious use is frozen in amber, and its role as a political symbol persists. Last year Lebanon's President Michel Sleiman gave Pope Benedict XVI an ivory-and-gold thurible. In 2007 Philippine President Gloria Macapagal-Arroyo gave an ivory Santo Niño to Pope Benedict XVI. For Christmas in 1987 President Ronald Reagan and Nancy Reagan bought an ivory Madonna originally presented to them as a state gift by Pope John Paul II. All these gifts made international headlines. Even Kenya's President Daniel arap Moi, father of the global ivory ban, once gave Pope John Paul II an elephant tusk. Moi would later make a bigger symbolic gesture, setting fire to thirteen tons of Kenyan ivory, perhaps the most iconic act in conservation history.

Father Jay is curator of his archdiocese's annual Santo Niño exhibition, which celebrates the best of his parishioners' collections and fills a two-story building outside Manila. The more than two hundred displays are drenched in so many fresh flowers and enveloped in such soft "Ave Maria" music that I'm reminded of a funeral as I look at the pale bodies dressed up like tiny kings. Ivory Santo Niños wear gold-plated crowns, jewels, and Swarovski crystal necklaces. Their eyes are hand-painted on glass imported from Germany. Their eyelashes are individual goat hairs. The gold thread in their capes is real, imported from India.

The elaborate displays are often owned by families of surprisingly modest means. Devotees have opened bankbooks in the names of their ivory icons. They name them in their wills. "I don't call it extravagant," Father Jay says. "I call it an offering to God." He surveys the child images, some of which are decorated in *lagang*, silvery mother of pearl flowers carved from nautilus shells. "When it comes to Santo Niño devotion," he says, "too much is not enough. As a priest, I've been praying, 'If all of this stuff is plain stupid, then God, put a stop to this.'"

Father Jay points to a Santo Niño holding a dove. "Most of the old ivories are heirlooms," he says. "The new ones are from Africa. They come in through the back door." In other words, they're smuggled. "It's like straightening up a crooked line: You buy the ivory, which came from a hazy origin, and you turn it into a spiritual item. See?" he says, with a giggle. His voice lowers to a whisper. "Because it's like buying a stolen item."

People should buy new ivory icons, he says, to avoid swindlers who use tea or even Coca-Cola to stain ivory to look antique. "I just tell them to buy the new ones, so the history of an image would start in you."

When I ask how new ivory gets to the Philippines, he tells me that Muslims from the southern island of Mindanao smuggle it in. Then, to signal a bribe, he puts two fingers into my shirt pocket. "To the coast guards, for example," he says. "Imagine from Africa to Europe and to the Philippines. How long is that kind of trip by boat?" He puts his fingers in my pocket again. "And you just keep on paying so many people so that it will enter your country."

It's part of one's sacrifice to the Santo Niño—smuggling elephant ivory as an act of devotion.

How to Smuggle Ivory

I had no illusions of linking Monsignor Garcia to any illegal activity, but when I told him I wanted an ivory Santo Niño, the man surprised me. "You will have to smuggle it to get it into the U.S."

"How?"

"Wrap it in old, stinky underwear and pour ketchup on it," he said. "So it looks shitty with blood. This is how it is done."

Garcia gave me the names of his favorite ivory carvers, all in Manila, along with advice on whom to go to for high volume, whose wife overcharges, who doesn't meet deadlines. He gave me phone numbers and locations. If I wanted to smuggle an icon that was too large to hide in my suitcase, I might get a certificate from the National Museum of the Philippines declaring my image to be antique, or I could get a carver to issue a paper declaring it to be imitation or alter the carving date to before the ivory ban. Whatever I decided to commission, Garcia promised to bless it for me. "Unlike those animal-nut priests who will not bless ivory," he said.

A few families control most of the ivory carving in Manila, moving like termites through massive quantities of tusks. Two of the main dealers are based in the city's religious-supplies district, Tayuman. During my five trips to the Philippines I visited every one of the ivory shops Garcia recommended to me and more, inquiring about buying ivory. More than once I was asked if I was a priest. In almost every shop someone proposed a way I could smuggle ivory to the U.S. One offered to paint my ivory with removable brown watercolor to resemble wood; another to make identical hand-painted statuettes out of resin to camouflage my ivory baby Jesus. If I was caught, I was told to lie and say "resin" to U.S. Customs. During one visit a dealer said Monsignor Garcia had just called and suggested that since I'd mentioned that my family had a funeral business, I might take her new, twenty-pound Santo Niño home by hiding it in the bottom of a casket. I said he must have been joking, but she didn't think so.

Priests, *balikbayans* (Filipinos living overseas), and gay Filipino men are major customers, according to Manila's most prominent ivory dealer. An antique dealer from New York City makes regular buying missions, as does a dealer from Mexico City, gathering up new ivory crucifixes, Madonnas, and baby Jesuses in bulk and smuggling them home in their luggage. Wherever there is a Filipino, I was often reminded, there is an altar to God.

And it seems Father Jay was right about a Muslim supply route. Several Manila dealers told me the primary suppliers are Filipino Muslims with connections to Africa. Malaysian Muslims figured into their network too. "Sometimes they bring it in bloody, and it smells bad," one dealer told me, pinching her nose.

Today's ivory trafficking follows ancient trade routes—accelerated by air travel, cell phones, and the Internet. Current photos I'd seen of ivory Coptic crosses on sale beside ivory Islamic prayer beads in Cairo's market now made more sense. Suddenly, recent ivory seizures on Zanzibar, an Islamic island off the coast of Tanzania—for centuries a global hub for trafficking slaves and ivory—seemed especially ominous, a sign that large-scale ivory crime might never go away. At least one shipment had been headed for Malaysia, where several multi-ton seizures were made last year.

The Philippines' ivory market is small compared with, say, China's, but it is centuries old and staggeringly obvious. Collectors and dealers share photographs of their ivories on Flickr and Facebook. CITES, as administrator of the 1989 global ivory ban, is the world's official organization standing between the slaughter of the 1980s—in which Africa is said to have lost half its elephants, more than 600,000 in just those ten years—and the extermination of the elephant. If CITES has overlooked the Philippines' ivory trade, what else has it missed?

In Praise of Pachyderms

THE ECONOMIST

Published in the *Economist* in 2017, "In Praise of Pachyderms" pro-vides some of the most recent data on elephant numbers and the effects of poaching on the endangerment of this iconic species. The situation is dire. African savanna elephants have lost close to a third of their numbers during a single decade. The pan-African "Great Elephant Census" of 2016 indicates only 350,000 African savanna elephants remain, which makes a total decline of around 140,000 since a previous census conducted in 2009. This article also suggests the unique importance of elephants as unusually intelligent and empathetic animals whose social systems may be second only to human societies in their complexity, making the loss of elephants a tragedy both for science and the human spirit. Such grim news is somewhat counterbalanced by the information that new methods of reducing human-elephant conflict are in the works, including the use of "bee fences" to discourage elephants from entering farms in Kenya. That's a hopeful possibility, while the genuinely positive information, we can read, is the closing of China's legal ivory markets. In recent times, luxury purchases of carved ivory in China have accounted for nearly three-quarters of the world total demand for ivory; the news of that legal change has already reduced end-market prices of ivory from a high of $2,100 per kilo in 2014 to a low of $730 per kilo in 2017. That indicates a spectacular drop in demand, which, one hopes, bodes well for the future of the

magnificent pachydermic animal whose gleaming front teeth are so
strangely prized by the human animal.

——— —— —— —— —— —— —— ——

T he symbol of the World Wide Fund for Nature is a giant panda.
The panda's black-and-white pelage certainly makes for a striking logo. But, though pandas are an endangered species, the cause
of their endangerment is depressingly quotidian: a loss of habitat as
Earth's human population increases. A better icon might be an elephant, particularly an African elephant, for elephants are not mere
collateral damage in humanity's relentless expansion. Often, rather,
they are deliberate targets, shot by poachers, who want their ivory;
by farmers, because of the damage they do to crops; and by cattle
herders, who see them as competitors for forage.

    In August 2016 the result of the Great Elephant Census, the most
extensive count of a wild species ever attempted, suggested that
about 350,000 African savannah elephants remain alive. This is down
by 140,000 since 2007. The census, conducted by a team led by Mike
Chase, an ecologist based in Botswana, and paid for by Paul Allen,
one of the founders of Microsoft, undertook almost 500,000 kilometres of aerial surveys to come to its conclusion—though the team
were unable to include forest elephants, a smaller, more reclusive
type that live in west and central Africa, and which many biologists
think a separate species.

    That most of the decline has been brought about by poaching is
scarcely in doubt. Seizures of smuggled ivory, and the size of the
carved-ivory market compared with the small amount of legal ivory
available, confirm it. But habitat loss is important, too—and not just
the conversion of bush into farmland. Roads, railways and fences,
built as Africa develops, stop elephants moving around. And an elephant needs a lot of room. According to George Wittemyer of Save

the Elephants (STE), a Kenyan research-and-conservation charity, an average elephant living in and around Samburu National Reserve, in northern Kenya, ranges over 1,500 square kilometres during the course of a year, and may travel as much as 60 kilometres a day.

The Long Road to Knowledge

The question, then, is whether elephants and people can ever coexist peacefully. And many of those who worry that the answer may be "no" fear the loss of more than just another species of charismatic megafauna. Elephants, about as unrelated to human beings as any mammal can be, seem nevertheless to have evolved intelligence and possibly even consciousness. Though they may not be alone in this (similar claims are made for certain whales, social carnivores and a few birds), they are certainly part of a small and select group. Losing even one example of how intelligence comes about and makes its living in the wild would not only be a shame in its own right, it would also diminish the ability of biologists of the future to understand the process, and thus how it happened to human beings.

Most of what is known about elephant society has been found out by STE's study in Samburu and by an even longer-running project, led by Cynthia Moss, at Amboseli National Park, in the country's south. Both use a mixture of good, old-fashioned fieldcraft and high-tech radio collars that permit individual animals to be tracked around by satellite.

Dr. Moss began her work in Amboseli in 1972, after collaborating in Tanzania with Iain Douglas-Hamilton, a zoologist who had been studying the animals since 1965. In 1993 Dr. Douglas-Hamilton, who had held various conservation-related jobs in the interim, followed suit by creating STE and recruiting Dr. Wittemyer to set up a research project in Samburu. That project now monitors seventy family groups comprising about three hundred adult females and their offspring, and also around two hundred adult males. Since they began work, Dr. Wittemyer and his team have collected more than 25,000 field observations of what the animals are up to, and around 4 million individual satellite locations.

Dr. Wittemyer argues that, human beings aside, no species on Earth has a more complex society than that of elephants. And elephant society does indeed have parallels with the way humans lived before the invention of agriculture.

The nuclei of their social arrangements are groups of four or five females and their young that are led by a matriarch who is mother, grandmother, great-grandmother, sister or aunt to most of them. Though males depart their natal group when maturity beckons at the age of twelve, females usually remain in it throughout their lives. Within a group, most adult females have, at any given moment, a single, dependent calf. They will not give birth again until this offspring is self-sufficient, which takes about four years. From a male point of view, sexually receptive females are therefore a rare commodity, to be sought out and often fought over. Such competition means that, though capable of fatherhood from the age of about fourteen, a male will be lucky to achieve it before he is in his twenties. Until that time arrives, he will be seen off by stronger rivals.

Were this all there was to elephant society, it would still be quite complex by mammalian standards—similar in scope to that of lions, which also live in matriarchal family groups that eject maturing males. But it would not deserve Dr. Wittemyer's accolade of near-human sophistication. Unlike lions, however, elephants have higher levels of organization, not immediately obvious to the observer, that are indeed quite humanlike.

First of all, families are part of wider "kinship" groups that come together and separate as the fancy takes them. Families commune with each other in this way about 10 percent of the time. On top of this, each kinship group is part of what Dr. Douglas-Hamilton, a Scot, calls a clan. Clans tend to gather in the dry season, when the amount of habitat capable of supporting elephants is restricted. Within a clan, relations are generally friendly. All clan members are known to one another and, since a clan will usually have at least one hundred adult members, and may have twice that, this means an adult (an adult female, at least) can recognize and have meaningful social relations with that many other individuals.

A figure of between one hundred and two hundred acquaintances is similar to the number of people with whom a human being can maintain a meaningful social relationship—a value known as Dunbar's number, after Robin Dunbar, the psychologist who proposed it. Dunbar's number for people is about 150. It is probably no coincidence that this reflects the maximum size of the human clans of those who make their living by hunting and gathering, and who spend most of their lives in smaller groups of relatives, separated from other clan members, scouring the landscape for food.

Dealing with so many peers, and remembering details of such large ranges, means elephants require enormous memories. Details of how their brains work are, beyond matters of basic anatomy, rather sketchy. But one thing which is known is that they have big hippocampuses. These structures, one in each cerebral hemisphere, are involved in the formation of long-term memories. Compared with the size of its brain, an elephant's hippocampuses are about 40 percent larger than those of a human being, suggesting that the old proverb about an elephant never forgetting may have a grain of truth in it.

À la recherche du temps perdu

In the field, the value of the memories thus stored increases with age. Matriarchs, usually the oldest elephant in a family group, know a lot. The studies in Amboseli and Samburu have shown that, in times of trouble such as a local drought, this knowledge permits them to lead their groups to other, richer pastures visited in the past. Though not actively taught (at least, as far as is known), such geographical information is passed down the generations by experience. Indeed, elephant biologists believe the ability of the young to benefit by and learn from the wisdom of the old is one of the most important reasons for the existence of groups—another thing elephants share with people.

Group living brings further advantages, as well—most notably those of collective defence. For, though most predators apart from humans armed with rifles would hesitate to attack an adult elephant, they will happily take on a youngster. A lone mother would be able to defend her calf against a single such predator, but many carnivores, particularly lions and hyenas, come in prides or packs. The solidarity of sisterhood means a group of elephants can usually deter attacks by its mere existence, and if deterrence does not work, then collective defence usually does. Here, again, experience seems to count. Data collected by Dr. Moss's team suggest that groups led by young matriarchs are more vulnerable to predation than those with older leaders.

Nor is it only in their social arrangements that elephants show signs of parallel evolution with humans. They also seem to have a capacity for solving problems by thinking about them in abstract terms. This is hard to demonstrate in the wild, for any evidence is necessarily anecdotal. But experiments conducted on domesticated Asian elephants (easier to deal with than African ones) show that they can use novel objects as tools to obtain out-of-reach food without trial and error beforehand. This is a trick some other species, such as great apes, can manage, but which most animals find impossible.

Wild elephants engage in one type of behavior in particular that leaves many observers unable to resist drawing human parallels. This is their reaction to their dead. Elephant corpses are centres of attraction for living elephants. They will visit them repeatedly, sniffing them with their trunks and rumbling as they do so. This is a species-specific response; elephants show no interest in the dead of any other type of animal. And they also react to elephant bones, as well as bodies, as Dr. Wittemyer has demonstrated. Prompted by the anecdotes of others, and his own observations that an elephant faced with such bones will often respond by scattering them, he laid out fields of bones in the bush. Wild elephants, he found, can distinguish their conspecifics' skeletal remains from those of other species. And they do, indeed, pick them up and fling them into the bush.

Elephants, then, are of great scientific curiosity. But, as its name suggests, Save the Elephants was not set up solely for the disinterested pursuit of knowledge. Indeed, as has often proved the way in field studies of other species, the focus of almost all elephant researchers, not just those in Kenya, has shifted from understanding the animals to preserving them.

Though poaching is still a threat in Kenya, changes in land use now seem an equal hazard. The human inhabitants of the area around the Samburu reserve (some of whom have given their tribal name to the place) have traditionally made their livings as pastoralists, driving herds of cattle from grazing place to grazing place. One source of conflict with elephants has been competition for pasture as the herders' populations have grown. Indeed, the reserve itself is now sometimes invaded by cowherds and their stock. But, on top of this, some pastoralists have begun to settle down. Buildings and fences are appearing on land which, though outside the reserve, is part of the local elephants' ranges as they travel from one place to another.

Here, the data Dr. Wittemyer and his team have accumulated can help. Satellite tracking that shows exactly how elephants move about can be used to steer decisions concerning land use in ways that help pachyderms. Elephants have places they prefer to live, which often correspond to protected areas, for the animals quickly work out where they are safe and where they are not. When travelling between these, which they usually do at night, they often follow narrow corridors.

Bee Off with You

Keeping such corridors clear of development is crucial to the well-being of the elephants which use them. Satellite maps are an important tool for doing so. Formal authorities in the country can take them into account, but, equally important, these maps are also quite persuasive in the public meetings at which local tribesmen agree on the use of what is collectively held land. Such meetings can assent to the legal "gazetting" of the corridors in question, to stop them being built on or fenced, so that elephants can pass freely.

This approach can work at a larger scale, as well. A new railway from Mombasa to Nairobi, for example, has been provided with elephant underpasses on routes used by the beasts—though an unintended consequence has been to encourage settlement near these transit points, which are useful for people, too. In the case of Samburu the satellite maps will be of great value if a proposed "development corridor," running inland from a planned expansion of the port of Lamu, goes ahead, as this may bring a new highway, railway and oil pipeline through land much used by elephants. Understanding elephants' behavior also permits them to be manipulated in ways that help reduce direct conflict between elephants and people. One such project harnesses elephants' fear of bee swarms.

Bees are the only animals apart from humans that elephants seem truly afraid of. Anecdotally, this has been known for a long time. But the matter has now been studied scientifically by Lucy King, a researcher at Oxford University who is also part of STE. Dr. King proved the anecdotes correct by playing the sound of a swarm of angry bees to wild elephants, and videoing the instant, panicked flight it provoked. The reason for this panic is that, although a bee's sting cannot penetrate most parts of an elephant's hide, swarms of bees tend to go for the eyes and the tip of the trunk, a pachyderm's most vulnerable parts. Bees are enemies that no amount of collective defense can discourage.

Armed with that knowledge, Dr. King and her colleague Fritz Vollrath came up with the idea of protecting farms with bee fences. The sort of fence most Kenyan smallholders can afford is too flimsy to exclude an elephant. But a bee fence, though flimsier still, does the job. It consists of pairs of poles about three meters apart, between which beehives can be hung like hammocks. The hives themselves are ten meters apart, and the poles are all connected by a single strand of wire 1.5 meters above the ground. This arrangement is enough to stop elephants in their tracks. Most are sufficiently wary of hives to avoid passing the fence in the first place—indeed, they are so wary that half of the hives can be cheap dummies, rather than the real thing, without reducing a fence's effectiveness. Those that do try to pass between the poles blunder into the wire and shake the adjacent hives, with predictable results, and rarely attempt a second passage.

Bee-fenced farms, Dr. King and Dr. Vollrath have discovered, suffer only a fifth as many elephant raids as those with conventional protection. As a bonus, the honey the bees produce is a useful source of revenue. Indeed, the fences are so successful that they are being tried out in at least a dozen other countries. Though it seems almost a Heath-Robinson solution to the problem, bee fencing may be an important part of reconciling the interests of elephants and people.

Jumbo Threat

All the bee fences in the world, however, will not help if the problem of poaching remains unsolved. And that, ultimately, means suppressing demand for ivory. For years this looked a fool's errand. Now, though, it does not, for good news has arrived from what many regard as an unexpected quarter: the government of China.

Though international trade in ivory is illegal, some countries permit internal sales—and do not always inquire too closely about where the tusks contributing to those sales have come from. In recent years China, which has permitted such sales, has been the world's largest ivory market, estimated to account for 70 percent of ivory sold. By the end of 2017, though, any sale of ivory in China will be illegal, and all licensed ivory dealers will have had to shut up shop. The Chinese do seem serious about this. Not only are dealers actually closing down, but an anti-ivory propaganda campaign has begun, with stars such as Yao Ming, a basketball player, and Li Bingbing, an actress, being recruited to shame those who continue to buy objects made from elephant tusks.

Though there is evidence of new workshops opening, and others expanding, in some of China's neighbours such as Vietnam, many people hope that China's ivory ban will prove a tipping point in the fight to preserve elephants. Already, the price of the stuff in China has come down by two-thirds, from a peak of $2,100 a kilogram in 2014 to $730 earlier this year. That is bad news for smugglers, and for the poachers who supply them. If the Chinese ban really does stick, rather than driving the trade underground, then it is just possible that historians of the future will record 2017 as having been the year of the elephant.

FICTIONAL
AND LITERARY
ELEPHANTS

Faithful Elephants

YUKIO TSUCHIYA

Yukio Tsuchiya's children's story *Faithful Elephants: A True Story of Animals, People, and War*, originally published in Japan in 1951, dramatizes a painfully tragic time at the Ueno Zoo in Tokyo near the end of World War II when enemy planes were daily flying overhead and dropping bombs.

Out of a fear that the walls and cages of the zoo would be breached and the freed animals cause havoc among the human population in Tokyo, the Japanese Army dictated that all the large and potentially dangerous animals at Ueno be put to death. Most of them—the big cats, big snakes, and bears—were poisoned, but the zoo's three elephants were too discerning to eat poisoned food and too thick-skinned to be injected. Instead, they were starved to death. It is a grim tale, given poignancy by a focus on the sensitivity, innocence, and individuality of the three elephants—John, Tonky, and Wanly—and the desperate empathy of the zookeepers forced to participate in withholding food from their beloved charges. After the elephants are dead, the zookeepers shake their fists at the bombers in the sky and cry out, "Stop the war! Stop the war! Stop all wars!"

Long regarded in Japan as a powerful antiwar story that might help build "strongholds of peace...in the hearts of adults and children when they realize the sorrow, misery, horror, and foolishness of war,"[49] *Faithful Elephants* has been published in more than a hundred editions

in Japan. It is read each year on the radio during the day set aside to
remember Japan's loss and surrender at the end of World War II.

――― ―― ―― ―― ―― ―― ―― ―― ――

The cherry blossoms are in full bloom at the Ueno Zoo. Their pet-
als are falling in the soft breeze and sparkling sun. Beneath the
cherry trees, crowds of people are pushing to enter the zoo on such
a beautiful day.

Two elephants are outside performing their tricks for a lively
audience. While blowing toy trumpets with their long trunks, the
elephants walk along large wooden logs.

Not far from the cheerful square, there stands a tombstone. Not
many notice this monument for the animals that have died at the
Ueno Zoo. It is quiet and peaceful here, and the sun warms every
corner.

One day, an employee of the zoo, while tenderly polishing the
stone, told me a sad story of three elephants buried there.

"Today," he said, "there are three elephants in this zoo. But years
ago, we had three different elephants here. Their names were John,
Tonky, and Wanly. At that time, Japan was at war. Gradually, the war
had become more and more severe. Bombs were dropped on Tokyo
every day and night, like falling rain.

"What would happen if bombs hit the zoo? If the cages were
broken and dangerous animals escaped to run wild through the city,
it would be terrible! Therefore, by command of the Army, all of the
lions, tigers, leopards, bears, and big snakes were poisoned to death.

"By and by, it came time for the three elephants to be killed. They
began with John. John loved potatoes, so the elephant keepers mixed
poisoned potatoes with the good ones when it was time to feed him.
John, however, was a very clever elephant. He ate the good potatoes,

but each time he brought a poisoned potato to his mouth with his trunk, he threw it to the ground, *kerplunk!*

"'As it seems there is no other way,' the zookeepers said, 'we must inject poison directly into his body.'

"A large syringe, the kind used to give shots to horses, was prepared. But John's skin was so tough that the big needles broke with a loud *snap*, one after the other. When this did not work, the keepers reluctantly decided to starve him to death. Poor John died seventeen days later.

"Then it was Tonky's and Wanly's turn to die. These two had always gazed at people with loving eyes. They were sweet and gentlehearted. The zookeepers wanted so much to keep Tonky and Wanly alive that they thought of sending them to the zoo in Sendai, far north of Tokyo.

"But what if bombs fell on Sendai? What if the elephants got loose and ran wild there? What would happen then?

"Tonky and Wanly, too, were doomed to be killed at the Ueno Zoo, just like all the other animals.

"The elephant keepers stopped feeding Tonky and Wanly. As the days passed, the elephants became thinner and thinner, weaker and weaker. Whenever a keeper walked by their cage, they would stand up, tottering, as if to beg, 'Give us something to eat. Please, give us water!' Their small, loving eyes began to look like round rubber balls in their drooping, shrunken faces. The once big, strong elephants had become a sad shape.

"All this while, the elephants' trainer loved them as if they were his own children. He could only pace in front of the cage and moan, 'You poor, poor, pitiful elephants!' One day Tonky and Wanly lifted their heavy bodies, staggered to their feet, and came close to their trainer. Squeezing out what little strength they had left, Tonky and Wanly made their appeal. They stood on their hind legs and lifted their front legs up as high as they could. Then, raising their trunks high in the air, they did their banzai trick. Surely their friend would reward them with food and water as he used to do.

"The trainer could stand it no longer. 'Oh, Tonky! Oh, Wanly!' he wailed, and dashed to the food shed. He carried food and pails of water and threw it at their feet. 'Here!' he said, sobbing, and clung to their thin legs. 'Eat your food! Please drink. Drink your water!'

"All of the other keepers pretended not to see what the trainer had done. No one said a word. The director of the zoo just sat very still, biting his lip and gazing at the top of his desk. No one was supposed to give the elephants any food. No one was supposed to give them any water. But everyone was hoping and praying that if the elephants could survive just one more day, the war might be over and the elephants would be saved.

"At last, Tonky and Wanly could no longer move. They just lay on their sides, hardly able to see the white clouds floating in the sky over the zoo. However, their eyes appeared clearer and more beautiful than ever.

"Seeing his beloved elephants dying this way, the elephant trainer felt as if his heart would break. He had no more courage to see them. All of the other keepers felt the same and they too stayed away from the elephants' cage.

"Over two weeks later, Tonky and Wanly were dead. Both died leaning against the bars of their cage with their trunks stretched high in the air, still trying to do their banzai trick for the people who once fed them.

"'The elephants are dead! They're dead!' screamed the elephant trainer as he ran into the office. He buried his head in his arms and cried, beating the desk top with his fist.

"The rest of the zookeepers ran to the elephants' cage and stumbled in. They took hold of Tonky's and Wanly's thin bodies, as if to shake them back to life. Everyone burst into tears, then stroked the elephants' legs and trunks in sorrow.

"Above them, in the bright blue sky, the angry roar of enemy planes returned. Bombs began to drop on Tokyo once more. Still clinging to the elephants, the zookeepers raised their fists to the sky and implored, 'Stop the war! Stop the war! Stop all wars!'

"Later, when the bodies of the elephants were examined, nothing was found in their washtub-like stomachs—not even one drop of water."

With tears in his eyes, the zookeeper finished his story. "These three elephants—John, Tonky, and Wanly—are now resting peacefully under this monument."

He was still patting the tombstone tenderly as the cherry blossoms fell on the graves, like snowflakes.

A Mahout and His War Elephant

VU HUNG

In the middle of the twentieth century, thousands of wild elephants roamed the forested mountains of central and southern Vietnam; that wild population supported a smaller population of tamed, trained, and working elephants. Villagers living in the highlands traditionally captured elephants by riding tamed and trained ones into the middle of wild herds. A mahout astride a trained elephant would carry a long stick projecting a kind of lasso—a rattan loop tied to a coil of buffalo-leather rope—and snag a young wild elephant by the foot. The snagged elephant would then be cornered and subdued by other mahouts riding other elephants and working with ropes, and in that manner a process began that ultimately, it was hoped, would result in an elephant who could work in logging, land-clearing, and hauling heavy loads. During the Vietnamese war of independence from the French colonialists after World War II, and then in the subsequent war of North Vietnam against the Americans and their allies in the south during the 1960s and early 1970s, those working elephants were used to transport military supplies through the mountains of central Vietnam. The war elephants of classical times were armored and armed and trained to support human warriors in close combat. The invention of gunpowder and other technological advances made classical war elephants obsolete, and the elephants of Vietnam today are on the edge of extinction; but it might be said that the ones used for transportation during the wars of the mid-twentieth century were also, in their own way, war elephants.

Vu Hung's *The Story of a Mahout and His War Elephant* is a novel for young readers based on a clear knowledge of the traditional methods for handling working elephants in Southeast Asia and portraying in broad strokes the coming-of-age story of a boy, Dik, and his young elephant, Lomluong, during a time of war. At the start of this tale, in 1942, Lomluong, still a wild elephant but separated from his herd and injured during a storm and earthquake, is captured by Dik and his grandfather, Rem, both of whom are riding on the back of the grandfather's own trained elephant, Lekdam. Old Rem is the sole guardian of his grandson, whose parents died during an epidemic, and Rem "blamed himself for his failure to provide property for his only grandson. What would happen to the boy if the legacy he inherited should consist solely of the ageing elephant?" Good fortune brought the young elephant Lomluong as a valuable legacy.

As Lomluong endures his taming and training to become a prime working elephant in the highland village of Takhan, so young Dik— accompanied by his best friend, Srung—passes through his own apprenticeship before he is handed the mahout's lance: an ironwood-handled, steel-tipped goad comparable to the traditional ankus of India or the bullhook of the West. Then, in 1945, the French, fighting to retain their colonial hold over the country, send in planes to strafe and bomb the village, leaving it in flames, with "many elephants...lying dead on the ground."

After the attack, all the villagers flee, taking the surviving elephants and retreating deeper into the highland jungles. Eventually they make contact with a group of communist guerrillas, and the villagers, having never before seen "revolutionary soldiers in the flesh," are amazed: the young warriors are "so brave and so simple," their guns such "outlandish weapons." Inspired by a visceral hatred of the French and a noble vision of fighting for independence, Dik and twenty other boys from the village join the guerrillas, trading their crossbows for guns and becoming soldiers of the Viet Minh force led by the legendary Ho Chi Minh. Dik also brings Lomluong, and the elephant is tasked with the duty of carrying weapons and ammunition for the revolutionary army. Fighting against the French brings the young mahout and his war elephant through a series of dangers and battles that help transform the boy into "a mature and seasoned combatant" and a hero who, after losing an arm in action, insists on staying with the revolutionary force until the French surrender in 1954. He then returns to his home village, marries, and becomes a father, but, as the book's epilogue warns, "it would not be long before the cruel lackeys of the American-Diem regime showed their faces in the village." Dik and Lomluong rejoin the army for the next fight against a new enemy.

Like the Japanese children's story *Faithful Elephants*, this war story presents the threat to elephants as a tragic side effect of human destructiveness. Elephants are innocent victims of war. Unlike the Japanese story, however, this one from Vietnam cannot be imagined as an antiwar polemic. It is more a recruiting pamphlet and propaganda poster in favor of war—the right and noble kind of war, that is. While dramatizing the virtues of heroic patriotism, it celebrates the social and psychological value of family and village and nation, the love between grandfather and grandson, and the bonds between comrades fighting for a just cause. How remarkable, then, that the bond most consistently celebrated in the book is that between a boy and an elephant, while the most intriguing virtue on display is that of kindness toward animals.

Of course, such kindness must be seen realistically within the larger context of the inevitable cruelty that occurs when a wild and free elephant is taken by force and transformed into a subdued and captive one. That larger context is as old as the history of human-elephant interactions, and the methods of taming and training are already familiar to us. Nevertheless, grandfather Rem, himself an honored and experienced mahout, teaches Dik the important principle that discipline alone is not sufficient for taming and training an elephant, that—as Rem declares in the following dramatic excerpt from *A Mahout and His War Elephant*—"Kindness is more effective than lance and chain put together. Animals, like human beings, resent harsh treatment."

It was still dark when Old Rem went out to see the village chief and to notify him of the event. At home, he got together the most valuable things he had—several bear's gall-bladders, a couple of deer antlers, a few monkey skeletons, and a number of fine pelts. He went down to the market with the lot, which he sold for silver coins. He needed cash to stand a treat to his fellow-mahouts and to pay the village chief and people in high places.

Hardly had he left when Srung turned up. The boy got up on the big stone mortar below the house and gave three light knocks on the underside of the floor. Above Dik was still sleeping soundly. He had stayed awake almost the whole night. Srung knocked again, harder this time. Dik woke up. Recognizing the signal he rushed down.

"I bet you don't know yet," he said, at once, giving his friend no time to speak. "Granddad captured a stray elephant last night."

"I knew. I heard the gongs. I heard the elephant. I knew everything."

"Granddad gave him to me. He's mine now."

The boys, hand in hand, skipped merrily to the park. It was not yet quite light. A few stars were blinking in the milky sky. The branches were riding a gentle breeze. Only the captured elephant remained motionless. It neither bellowed nor struggled, tired as it was.

The boys gazed at it in wonder. In the half-light they tried to make out its features.

"I don't know what's wrong with my stomach." Dik told his friend. "I didn't sleep a wink the whole night, and I couldn't eat anything either, not even the meat Granddad roasted for me!"

"That's because you're happy. You feel queer when you're happy."

"And you? Aren't you happy, too?"

"Of course, I am happy even though the elephant isn't mine."

It was true that Srung was not envious, although his family had no elephant, not even an old one like Lekdam. To the villagers, not to have an elephant was a disadvantage, even a grave misfortune.

Dik was moved by the thought, and he felt his joy selfish. Srung had no grandfather to look after him. A loving grandfather meant a lot.

"I'll ask Granddad to let you take care of the elephant with me," Dik offered generously, wrapping his arms about his friend's shoulders. "When we grow up we'll share him too. We'll let him join the herd in the village, and we'll even get one for you."

Now began the taming of the wild elephant.

Old Rem and the boys would ride Lekdam to the place where the captive was chained, and the boys would drop at Lomluong's feet shreds of tender banana stems and bundles of fragrant grass.

At first Lomluong just ignored the food. Then, unable to resist the pangs of hunger, it picked up some grass and slowly chewed it.

The old man and the boys meanwhile were singing high praises of their newly acquired treasure. What a magnificent calf. It was flawless. A high, stubborn forehead; a pair of intelligent eyes; a neat, delicate head. The tusks, hefty and smooth, were like two sappy bamboo shoots. The trunk, articulate, was as deft as a human hand. The feet, nimble though big, promised great endurance. "He'll be much better even than Lekdam in his youth," the old man chuckled.

Each feeding took several hours, but Old Rem and the boys had no objection to this: they could never admire the elephant too much. Then, the old mahout would order Lekdam to close in and lock the young elephant's rebellious trunk in its own, while Dik and Srung dismounted to administer balm to the wound on Lomluong's leg.

The wound having healed, a chain was put on Lomluong's hindleg. The elephant was marched by Old Rem and some other mahouts to the clearing and chained to a stake, where it stood swaying languidly waiting for food. But its mood would sometimes change. There were times when he would not touch anything, but just remained as motionless as a rock. Perhaps it was thinking of the old herd and the old days in freedom. It looked forlornly in the direction of the jungle and gave long, sad complaining brays. The boys were greatly alarmed, because it had been known that elephants had actually died of grief.

"There's no cause for fear," Old Rem would assure them. "Lomluong is young. Unlike old elephants, he will get used to us."

The elephant would strain its trunk and sniff at the stake. No scents of the wilderness. Only human scents, mostly the scents of human sweat, everywhere, on everything—the stake, the chain, the foot-prints on the ground. At this the elephant would become mad. It lashed at the chain. It lifted its trunk threateningly at people; it hissed and grunted.

It was still a beast of the wild.

One late afternoon at mealtime, heavy footfalls were heard coming from the clearing.

"Elephant gone wild!" Someone was shouting.

Dik and Srung rushed out. They were thoroughly scared by the sight that met their eyes. Lomluong had broken his chain and was giving chase to Old Rem. The trunk curled up, the eyes blood-shot in the setting sun, the elephant galloped madly behind the old man. Its heavy feet raised big clouds of dust.

Old Rem weaved his way among the trees for some time. Then, cutting a straight line, he made for an open place. His intention was obvious—to get to the old elephant, which was turning round on its chain in a frantic attempt to come to the rescue of its master.

Lomluong was gaining fast on the old man. Several times it threw out its trunk but somehow the old man managed to dodge it.

Meanwhile Dik had run home for a lance. Now he tried to cut in between the mad elephant and his grandfather. The elephant, caring little for the tiny human being and his frail weapon, moved on, irresistible as a rolling rock.

Looking on, Srung did not know what to do. Then the light from a nearby fire gave him an idea. He made a sprint for the fire and returned with a flaming brand which he flung at the head of the oncoming elephant, just in time to save the old man from its trunk.

Alarmed, the elephant made a sudden sideways lunge, which almost threw it off balance.

A dozen mahouts meanwhile had arrived with their elephants. One made his animal kneel down to receive Old Rem and the two boys, while the others set out after the rebel who had changed course and was now heading for the jungle.

The old man, breathing heavily, congratulated the boys on their courage. Then, receiving a lance from the leading mahout, he took his place in front and prodded the elephant into a quick gallop with the rest of the party.

Soon the rebel was stopped and surrounded. A dozen trunks reached out for it. In a jiffy, its trunk and legs were caught and locked.

The punishment began.

One mahout, riding a powerful bull, came forward. "Savage beast!" he roared, his eyes ablaze with wrath. "How dare you turn against your master?"

"Give it to him, and hard," he ordered. His mount, equally angry, began flailing the victim's chest with its enormous trunk.

A second mahout took his turn and was followed by a third. Lomluong took his punishment with resignation.

The animal's stoic attitude softened Old Rem.

"Stop, please," he pleaded, holding up his lance for attention.

The mahout who was directing his own share of the punishment would hear none of this. He urged his elephant to go on.

"Please, that's enough," Dik added his plea to his grandfather's.

"Yes, that's enough," Old Rem echoed again.

Srung was on the verge of tears. He looked around for sympathy. But the mahout was adamant.

"Don't you see, you had a narrow escape?" someone insisted to Old Rem.

"Let's beat this savage habit out of it," another added.

"It should not be allowed to set a bad example to the other elephants."

"This only serves it right."

Blows meanwhile continued to rain on Lomluong, which began groaning softly.

Dik was weeping bitterly now. "Stop, please," he sobbed. "He's badly hurt."

The sight of the elephant and the boys was too much for Old Rem. An experienced mahout, he knew that excessive severity would not do any good to an elephant which could take just punishment without flinching.

"Enough's enough, friends," he said firmly.

"Come, let's stop for the sake of Rem and the children," another elderly mahout intervened.

The mahout who was punishing Lomluong brought his lance sharply down on the neck of his mount. "Stop!" he snapped.

Dik and Srung were heartily relieved. They jumped down to the ground, ran away, and returned a moment later with a solid chain.

Old Rem also dismounted to help the boys put the chain on Lomluong, which was then marched back to the clearing and secured again to the stake.

Before Lomluong was left to itself, Dik stepped forward.

"A moment, please," he told the mahout whose elephant was keeping a hold on Lomluong's trunk.

Dik tiptoed up to his elephant, and spoke softly into its ear, "Don't ever be naughty again, Lomluong. Can't you see that we are your friends and are good to you?"

And he ran his hands soothingly over the bruised chest of the animal.

A few days later, after Lomluong had returned to its normal docile and dignified self, the old man and the boys began trying to win its confidence. They spent whole days with it, feeding it, calling its name. Sometimes they formed a circle around it, singing an old song, Old Rem mumbling the words in his throat, the boys shouting at the top of their voices.

> Little elephant, my little elephant,
> Don't ever get mad,
> Never run away,
> Stay and let's make friends.
> Mornings we give you food
> Maize and young shoots,
> Grass and roots,
> Leaves and fruit.
> Noons we take you home,
> Into the shade, into the coolness,
> Where you can play,
> Where you can rest.

The young elephant showed great impatience at first. It stamped its feet, jerked its trunk, or grunted menacingly. Like all other elephants in the early stage of captivity, it had an innate dislike of human scents and human voices.

The old man and the boys, however, were not easily discouraged. They kept walking around the elephant, feeding it, calling its name, singing.

> Little elephant, my little elephant,
> Don't return to the mountains.
> Never go away.
> Stay and play with us...

Days grew into months. The elephant could not beat Old Rem and the boys for patience. Moreover, time had a soothing effect on it. It began to flap its ears whenever its name was mentioned.

Old Rem decided it was time to break the elephant into human scents.

With Dik and Srung keeping watch from Lekdam's back, their lances at the ready, Old Rem approached Lomluong, a bucket of salted water in one hand, a lance in the other, his eyes fixed on those of the elephant.

Lomluong pulled back and lifted his trunk. Onlookers, including Dik and Srung on Lekdam and a few mahouts who happened to be around, held their breath.

The old man ordered Lomluong to stay still while he made several cautious steps forward. The elephant, its trunk still poised, began showing signs of uneasiness, and when the old man pricked the trunk in a very swift movement, it emitted a short grunt and drew back with a look of alarm.

Old Rem set the bucket down.

"Drink," he ordered. "It's good for you."

The elephant plunged its trunk deep into the bucket and took a quick draught. Then, as quickly, it threw up its trunk unaccustomed to the taste of salt. The second sip, however, was taken with some appreciation, and the elephant squinted quizzically at the old man.

Soon Lomluong was drinking with gusto, playfully waving its trunk at the old man after each gulp.

After several such sessions, when he was assured of the elephant's disposition to be friendly, the old man allowed the children to have a try. The latter, armed with lances, brought the bucket close to the elephant.

They saw the elephant put the tip of its trunk into its mouth, and knew what it was up to.

"Don't be naughty," Dik shrieked with delight when the elephant started spraying them.

This was just for fun. Lomluong had already accepted the children's friendship. It let down its trunk and waited for them, body swaying, eyes blinking. "Come here, brave boys. Come to me," it seemed to be saying. When the children came within reach, it snatched the bucket from their hands and drank eagerly to their great amusement.

Its drink finished, the elephant sniffed at the boys. The children were beside themselves with joy. They caressed the friendly trunk that was licking at the sweat on their bodies.

Plans were then made to train the elephant to work.

"Do we have to burn holes behind his ears, must we do it?" the children asked with foreboding. "He has such splendid ears."

"No, we don't have to. Lomluong deserves better than that," Old Rem reassured them.

A common practice among mahouts was to burn two holes—more precisely to inflict two permanent wounds—behind an elephant's ears to make it sensitive to punishment.

But Old Rem had his own method. He had cured the ear wounds on Lekdam, in spite of warnings by fellow-mahouts. He loved his elephant. He would not hurt it.

"What if he goes wild?" the children asked again.

"Listen, boys," Old Rem reasoned. "A mahout who only knows where to plunge his lance is not a good mahout. Kindness is more effective than lance and chain put together. Animals, like human beings, resent harsh treatment. Remember the story of the elephant which took revenge on its master?"

Of course the children remembered the story, having heard it so many times. The old man could also remember every word of it, and could still recall the conversation taking place many years ago between him and the Lao mahout who had first told the story to him.

It was thus: "There was a young mahout of great skill. He changed elephants frequently, before he finally found one to his liking. It was a very intelligent bull. The animal was very efficient, and the mahout made a lot of money with it. To do him justice, he was not unkind to his elephant. But he took to drinking, and you can't expect a drunken man to be kind and just. So, every time he got drunk he beat and cursed his elephant. As his drinking bouts became more frequent, there was no end to the elephant's misery. Finally he began to develop an unreasonable hatred of the animal. One day he made it haul a load of timber too heavy for its strength, however the animal succeeded in hauling it."

"What would you have done yourself?" the Lao mahout had asked young Rem at this point.

"I'd have rewarded the elephant. I'd have given him an armful of sugar cane or a bucket of rice soup."

"The same with me," the Lao mahout said approvingly. "We should never forget the services of our elephants."

"I would almost say the great favours they do us," Rem corrected.

"You're right," the storyteller conceded with a smile. "Now, this mahout, instead of rewarding his elephant, went out for a drink, leaving the animal alone, without food. The next time he went to it, the elephant gave him a warning blow with its trunk. The mahout should have understood the danger sign. When your elephant doesn't love you any more, that means danger. But, instead of taking the hint... and beginning to undo the wrongs he had been doing, the mahout re-opened the wounds behind the elephant's ears, into which he would viciously dig his lance every time the elephant disobeyed his unreasonable orders. Even animals prefer reason, don't they? So, one day, after it had been unchained the elephant gave its master one long look and roared angrily. The mahout did not expect this. "What are you up to? Seeking trouble?" he drawled drunkenly, and reached behind his back for his lance. But the lance was not there, and his face went white. He wheeled about and ran. Too late. The elephant had flung out its trunk and lunged forward. The man was caught around his waist, hurled up into the air, and impaled on the upraised tusks as he fell down. His body was never found. The elephant had run with it into the jungle."

Old Rem believed the story was no more than a fable. A thoroughly tamed elephant would never kill anyone, much less its master. But all the same he liked the story for its moral, and had all his life scrupulously acted on the advice derived from it. Now he wanted to drive home to the children the truth that training without kindness would achieve nothing.

So it was decided that no ear wound would be inflicted on Lomluong.

To make the elephant used to riders, loads of gradually increasing weights were put on its back. Lomluong balked at first, trying to throw everything off. But by dint of patience the old man and his aides succeeded in breaking the animal into this habit. Then they began taking their places on its neck.

The next step was much more difficult and exhausting. The elephant was taken to the clearing very early in the morning before it had eaten, for a full stomach would induce sluggishness of both body and mind.

To make it kneel and get up on orders, ropes were attached to its trunk and forelegs. The loose ends of the ropes were held by Dik and

Srung sitting on the animal's back together with the old man. The free space on the howdah was filled with sugar cane.

"Khuc Khau," the old man would say, and Dik would pull the leg ropes. This was repeated several times before Lomluong knew what was expected of it. At the next order, it would bend first its forelegs and then its hindlegs, learning to crouch down. The old man would then jump to the ground and throw some sugar cane to the elephant.

"Nhun Khun" was the order to get up. When this was given, Srung would give a slight tug to the trunk rope. The elephant caught the hint instantly. It stood up, first on its hindlegs, then on its forelegs. It was rewarded with more sugar cane.

Lomluong showed great intelligence. He was very quick at grasping the meanings of "Khuc Khau" and "Nhun Khun" and soon could execute the orders without the prompting of the ropes. Old Rem, for his part, always gave it just and reasonable treatment—scoldings for its mistakes, but generous rewards for progress.

For the lessons on steering, ropes were attached to the elephant's ears. At the command "Buong Xai" Dik would give a jerk to the left rope; Srung on the other side would pull the right one when the old man shouted, "Bung Khoa." The elephant learnt to react correctly to the signals, without much difficulty.

Lomluong developed a special liking for these outings, during which it could fare sumptuously and was free of all fetters. It was evidently bored by inactivity, and to show this it would rattle the chain fussily, or when food was brought in, would flap its ears, wag its trunk, and stamp its feet. "Hey, little friends. What about some fun now," it seemed to say.

The actual domestication of Lomluong had begun the first summer of his captivity when cuckoos, coming for the ripening litchi, were filling the jungle with their musical reiteration. It had continued into the cold months and into the next spring. By the time the noisy cuckoos were back again, Lomluong was completely tamed. It could carry out all orders without prompting, caring little for rewards.

Come winter, Lomluong would join the other working elephants of the village. Old Rem had prepared for it two solid chains and a thick breast-pad made of canvas. He also put Dik and Srung directly in charge of its health and well-being.

He entertained great hopes of his young favourite.

And Lomluong, for its part, had become very much attached to its masters, particularly Dik, who regarded it as a dear friend.

Everybody was looking forward to the day when Lomluong would be granted the status of a full grown work elephant.

Dear Elephant, Sir

ROMAIN GARY

I am ending this collection with a cri de coeur, a brilliantly imaginative essay written by the famed twentieth-century French aviator, diplomat, filmmaker, and novelist Romain Gary and published in a December 1967 issue of *Life*. Gary's letter to an elephant is simply the best argument I have seen that, after identifying the devastation inflicted by thousands of years of hunting for meat and ivory followed by "civilized man's" absurd folly of hunting for sport, embraces the idea of protecting elephants not as a pragmatic act resulting from scientific or rational analysis, but as an intuitively moral, indeed spiritual act. We must make peace with elephants, Gary writes, if only because these magnificent and mysterious creatures impart "a resonance that cannot be accounted for in terms of science or reason, but only in terms of awe, wonder and reverence."

———— —— —— —— —— —— —— —— ——

Dear Elephant, Sir:
You will probably wonder, reading this letter, what could have prompted a zoological specimen so deeply preoccupied with the future of his own species, to write it. The reason of course, is self-preservation. For a long time now I have had the feeling that our destinies are linked. In these perilous days of "the balance of terror," of overkill and estimates telling how many of us could hope to survive a nuclear holocaust, it is only natural that my thoughts should turn

to you. In my eyes, dear Elephant, sir, you represent to perfection
everything that is threatened today with extinction in the name of
progress, efficiency, ideology, materialism, or even reason, for a certain abstract, inhuman use of reason and logic are becoming more
and more allies of our murderous folly. It seems clear today that we
have been merely doing to other species, and to yours in the first
place, what we are on the verge of doing to ourselves.

We met for the first time almost a century ago in my nursery.
We shared the same bed for many years, and I never went to sleep
without kissing your trunk and then holding you tightly in my arms,
until the day came when my mother took you away, telling me, with a
certain absence of logic, that I was a big boy now and therefore could
no longer have an elephant for a pet. Psychologists will no doubt say
that my "fixation" on elephants goes back to that painful moment of
separation, and that my longing for your company is actually a nos-
talgia for my long gone innocence and childhood. And, indeed, you
are precisely that in my eyes: a symbol of purity, a dream of paradise
lost, a yearning for the impossible, of man and beast living peacefully
together.

Years later, somewhere in Sudan, we met again. I was returning
from a bombing mission over Ethiopia, and brought down my dam-
aged plane south of Khartoum, on the western bank of the Nile. I
walked for three days to reach water and to have the most satisfying
drink of my life, thus, as it turned out later, contracting typhoid and
almost dying. You appeared before me among some meager acacia
trees, and at first I thought I was a victim of hallucination. For you
were red, dark red, from trunk to tail, and the sight of a red elephant
sitting on his rear end and purring made my hair stand on end. Yes,
you were purring. I have learned since then that this deep rumbling
is a sign of satisfaction, and I suppose the bark of the tree you were
eating was particularly delicious. It took me some time to realize that

you were red from wallowing in the mud and that meant the proximity of water. I edged forward, and you became aware of my presence. You perked up your ears, and your head seemed to triple in size, while your whole mountain of a body disappeared behind those suddenly hoisted sails. You were no more than sixty or seventy yards from me and I could not only see your eyes but feel them, as if my stomach had eyes of its own. I was too weak to run. Besides, my exhaustion, fever and thirst were greater than my fear. I therefore did the only thing that I could do under the circumstances: I gave up. I have given up quite a few times, during the war, closing my eyes and waiting for death, and each time I have been given a medal for bravery.

When I opened my eyes again, you were asleep. I suppose you had not seen me or had taken one look at me and became overcome with boredom. Anyway, you were standing there, your trunk limp, your ears collapsed, your eyes closed, and I remember that tears came to my eyes. I was seized by an almost irresistible urge to come close to you, to press your trunk against me, to huddle against your hide, and there, fully protected, to sleep peacefully forever. The strangest feeling came over me: I knew it was my mother who had sent you. She had relented at last and had given you back to me.

I took a step in your direction, then another.... For a man as utterly tired as I was, there was something strangely reassuring about your huge, rocklike sight. I knew that if I could touch you, caress you, lean against you, you would give me some of your life force. It was one of those moments when a man needs so much energy and so much strength to overcome and to prevail that he thinks of God. I have never been able to raise my eyes that high, and so I stop at elephants.

I was quite close to you when I stumbled and fell. And then it happened. The earth shook under me and the most terrifying sound of a thousand donkeys braying at once with a lion's voice turned my heart into a captive grasshopper. As a matter of fact, I screamed too, and my yell had all the frightening strength of a two-month-old baby. The next thing I knew I was running like a champion rabbit across the clearing, yelling at the top of my lungs, and it seemed that some of your strength had indeed been transmitted to me, for never has a half-dead man come to life quicker and run faster. As a matter of fact, we both were running, although in opposite directions. You were trumpeting away and I was shrieking away with the voice of blind fear, and as I needed all my energy and could not waste any of it on controlling *all* of my muscles... but the least said about that the better. Besides, one has to pay something for one's bravery. After all, I had scared an elephant.

We never met again, and yet in our thwarted, restricted, controlled, indexed and repressed existence, the echo of your irrepressible thundering march through the open spaces of Africa keeps reaching me, awakening a confused longing. It sounds triumphantly like the end of acceptance and servitude, an echo of limitless freedom that has haunted our soul since the beginning of time.

I hope you won't consider me discourteous if I tell you that your size, strength and craving for unrestricted existence make you quite obviously anachronistic. You're therefore considered as incompatible with modern times, and for all of us who are sick and tired of our polluted cities and even more polluted minds, your colossal presence and the fact of your survival against all odds acts as a God-sent reassurance. Everything is not yet lost, the last hope of freedom has not yet vanished completely from this earth and, who knows, if we stop destroying elephants and save them from extinction, we may yet succeed in protecting our own species from our destructive enterprises as well.

If Man shows himself capable of respect for life in its hugest and most cumbersome form—now, now, don't flip your ears and raise your trunk angrily, no insult intended—then a chance remains that China is not pointing the way to our future and that the other cumbersome, clumsy prehistoric monster, individual man, will somehow manage to survive.

Years ago, I met a Frenchman who had devoted himself body and soul to the defense of the African elephant. Somewhere within the rolling green sea of what was known as Tchad territory, under the stars that always seem to shine brighter when a man's voice manages to rise higher than his solitude, he told me: "Dogs are not enough. People never felt more lost, more lonely in this man-made world. They need company, a stronger, bigger company than ever. Something that can really stand up to it all. Dogs aren't enough, what we need is elephants."

And who knows? We may even need a companion infinitely bigger and more powerful than that.

I can almost see an ironic twinkle in your eyes as you read my letter. And no doubt you prick your ears, deeply mistrustful of every human sound. Have they ever told you that your ear has almost exactly the shape of the African continent? Your gray, rocklike mass has the very color and texture of Mother Earth herself. There is something incongruous and almost girlish about your eyelashes, and your rump is that of a monstrous puppy.

For thousands of years you have been hunted for meat and ivory, but it is civilized man who has invented killing you for pleasure and trophy. Everything that is frightened, frustrated, weak and insecure in us seems to find a sick comfort in killing the most powerful of all earth's creatures. This wanton act brings the kind of "virile" reassurance that casts a strange light on the nature of our virility.

There are those, of course, who say you are useless, that you destroy crops in a land where starvation is rampant, that mankind has enough problems taking care of itself, without being expected to burden itself with elephants. They are saying, in fact, that you are luxury, that we can no longer afford you. This is exactly the kind of argument every totalitarian regime from Stalin and Hitler to Mao uses to prove that a truly "progressive" society cannot be expected to afford the luxury of individual freedom. Human rights are elephants, too. The right of dissent, of independent thinking, the right to oppose and to challenge authority can very easily be throttled and repressed in the name of "necessity."

In a German prison camp, during the last world war, you played, Elephant, sir, a lifesaving role. Locked behind the barbed wires we would think of the elephant herds thundering across the endless plains of Africa, and the image of such an irresistible liberty helped us to survive. If the world can no longer afford the luxury of natural beauty, then it will soon be overcome and destroyed by its own ugliness. I myself feel deeply that the fate of Man, and his dignity, are at stake whenever the earth's natural splendors are threatened with extinction.

The task of remaining human seems at times almost overwhelming. And yet it is essential that we should shoulder on our backbreaking walk toward the unknown a supplementary burden: the burden of elephants. There is no doubt that in the name of total rationalism you should be destroyed, leaving all the room to us on this overpopulated planet. Neither can there be any doubt that your disappearance will mean the beginning of an entirely man-made world. But let me tell you this, old friend: in an entirely man-made world, there can be no room for man either. All that will be left of us are robots. We are not and could never be our own creation. We are forever condemned to be part of a mystery that neither logic nor imagination can fathom, and your presence among us carries a resonance that cannot be accounted for in terms of science or reason, but only in terms of awe, wonder and reverence. You are our last innocence....

Your very devoted friend,
Romain Gary

BIBLIOGRAPHY

Anthony, Lawrence, with Graham Spence. *The Elephant Whisperer: My Life with the Herd in the African Wild*. New York: St. Martin's, 2009.

Aristotle. *Historia Animalium*, volume 1, translated by A. L. Peck. Loeb Classical Library Volume 437. Cambridge, Mass.: Harvard University Press, 1965.

Aristotle. *Historia Animalium*, volume 2, translated by A. L. Peck. Loeb Classical Library Volume 439. Cambridge, Mass.: Harvard University Press, 1991.

Aristotle. *Historia Animalium*, volume 3, translated by D. M. Balme. Loeb Classical Library Volume 437. Cambridge, Mass.: Harvard University Press, 1965.

Arrian. *Alexander the Great: The Anabasis and the Indica*. Translated by Martin Hammond. Oxford: Oxford University Press, 2013.

Barnum, Phineas T. "Jumbo the Elephant." In *The Colossal P. T. Barnum Reader: Nothing Else Like It in the Universe*, edited by James W. Cook. Chicago: University of Illinois Press, 2005.

Bell, W. D. M. *The Wanderings of an Elephant Hunter*. New York: Charles Scribner's Sons, 1923.

Bradley, Carol. *Last Chain on Billie: How One Extraordinary Elephant Escaped the Big Top*. New York: St. Martin's, 2014.

Christy, Bryan. "Ivory Worship." *National Geographic*, October 2012.

Cosentino, Donald J. "The Talking (Gray) Heads: Elephant as Metaphor in African Myth and Folklore." In *Elephant: The Animal and Its Ivory in African Culture*, edited by Doran H. Ross. Los Angeles: Fowler Museum of Cultural History, University of California, 1992.

Douglas-Hamilton, Iain, and Oria Douglas-Hamilton. *Among the Elephants*. New York: Penguin Books, 1975.

Edgerton, Franklin, trans. *The Elephant-Lore of the Hindus*. Delhi: Motilal Banarsidass, 1985.

Gale, U Toke. *Burmese Timber Elephant.* Rangoon: Trade Corporation 9, 1974.

Gary, Romain. "Dear Elephant, Sir." *Life,* December 22, 1967.

Gordon-Cumming, Roualeyn. *Five Years of a Hunter's Life in the Far Interior of South Africa.* Vol. 2. New York: Harper & Brothers, 1850.

Hung, Vu. *The Story of a Mahout and His War Elephant.* Hanoi: Foreign Languages Publishing House, 1976.

"In Praise of Pachyderms." *Economist,* June 17, 2017.

Knappert, Jan. *Myths and Legends of the Congo.* London: Heinemann, 1971.

McComb, Karen, Lucy Baker, and Cynthia Moss. "African Elephants Show High Levels of Interest in the Skulls and Ivory of Their Own Species." *Biology Letters* 2, no. 1 (2006).

Moss, Cynthia. *Elephant Memories: Thirteen Years in the Life of an Elephant Family.* Chicago: University of Chicago Press, 2000.

Payne, Katy. *Silent Thunder: In the Presence of Elephants.* New York: Simon & Schuster, 1998.

Payne, Katy. "Sources of Social Complexity in the Three Elephant Species." In *Animal Social Complexity: Intelligence, Culture, and Individualized Societies,* edited by Frans B. M. de Waal and Peter L. Tyack. Cambridge, Mass.: Harvard University Press, 2003.

Pliny the Elder. *Natural History: A Selection,* translated by John F. Healy. New York: Penguin Books, 1991.

Plotnik, Joshua M., Frans B. M. de Waal, and Diana Reiss. "Self-Recognition in an Asian Elephant." *PNAS* 103, no. 45 (2006).

Poole, Joyce. *Coming of Age with Elephants.* New York: Hyperion, 1996.

Price, Charles Edwin. *The Day They Hung the Elephant.* Johnson City, Tenn.: Overmountain Press, 1992.

Safina, Carl. "Big Love: The Emotional Lives of Elephants." *Orion,* May/June 2015.

Tsuchiya, Yukio. *Faithful Elephants: A True Story of Animals, People and War.* Translated by Tomoko Tsuchiya Dykes. Boston: Houghton Mifflin, 1988.

Turnbull, Colin M. "Legends of the BaMbuti." *Journal of the Royal Anthropological Institute of Great Britain and Ireland* 89, no. 1 (1959).

Watson, Lyall. *Elephantoms: Tracking the Elephant.* New York: W. W. Norton, 2000.

White, T. H., ed. and trans. *The Book of Beasts: Being a Translation from a Latin Bestiary of the Twelfth Century.* New York: Dover, 1984.

Williams, J. H. *Elephant Bill.* N.p.: Long Riders' Guild Press, 2001.

ACKNOWLEDGMENTS AND CREDITS

I am grateful to Barbara Ras of Trinity University Press, who originally described the idea for this book and asked me to create it as the editor. Marguerite Avery at Trinity intelligently guided me to revise and reshape the material into its current form, and for her support I am equally grateful. Further thanks are due to those friends, allies, and colleagues who encouraged me in various ways to think broadly and creatively about elephants and the literature of elephantology. They include Marc Bekoff, Daniel C. Dennett, Shubhobroto Ghosh, Wyn Kelley, Felicia Nutter, Karen Panetta, and Allen Rutberg. I must also give a special thanks to the team of Trinity interns: Isaiah Mitchell, Georgie Riggs, Ariana Fletcher-Bai, Allison Carr, and Erica Schoenberg.

Excerpt from *Alexander the Great: The Anabasis and the Indica*, by Arrian, translated by Martin Hammond. Reprinted by permission of Oxford University Press.

Excerpt from *Historia Animalium*, vol. 1, by Aristotle. Vol. IX, translated by A. L. Peck, Loeb Classical Library Volume 437. Cambridge, MA: Harvard University Press. Copyright © 1965 by the President and Fellows of Harvard College. Loeb Classical Library ® is a registered trademark of the President and Fellows of Harvard College.

Excerpt from *Historia Animalium*, vol. 2, by Aristotle. Vol. X, translated by A. L. Peck, Loeb Classical Library Volume 438. Cambridge, MA: Harvard University Press. Copyright © 1970 by the President and Fellows of Harvard College. Loeb Classical Library ® is a registered trademark of the President and Fellows of Harvard College.

Excerpt from *Historia Animalium*, vol. 3, by Aristotle. Vol. XI, translated by D. M. Balme, Loeb Classical Library Volume 439. Cambridge, MA: Harvard University Press. Copyright © 1991 by the President and Fellows of Harvard College. Loeb Classical Library ® is a registered trademark of the President and Fellows of Harvard College.

Excerpt from *Natural History: A Selection*, by Pliny the Elder, translated and with an introduction and notes by John F. Healy. Copyright © 1991 by John F. Healy. Reprinted by permission of Penguin Books.

Excerpt from *Burmese Timber Elephant*, by U Toke Gale. Rangoon: Trade Corporation, 1974. Reprinted courtesy of the publisher.

Excerpt from *Elephant Bill: The Best-Selling Account of 1920s Life in the Jungles of Burma*, by J. H. Williams. Long Riders' Guild Press, 2001. Printed with permission of the publisher.

Excerpt from *The Day They Hung the Elephant*. Copyright © 1992 by Charles Edwin Price. Reprinted courtesy of Overmountain Press.

Excerpt from *Last Chain on Billie: How One Extraordinary Elephant Escaped the Big Top*. Copyright © 2014 by Carol Bradley. Reprinted by permission of St. Martin's Griffin, an imprint of St. Martin's Press. All rights reserved.

"Abusing Captive Elephants in India," by Shubhobroto Ghosh. Reprinted by permission of the author.

Excerpts from Iain and Oria Douglas-Hamilton, *Among the Elephants*. Penguin Books, 1975. Reprinted with permission of the publisher.

Excerpts from *Elephant Memories: Thirteen Years in the Life of an Elephant Family*, by Cynthia Moss. Copyright © 2000 by Cynthia Moss. Reprinted with permission.

Excerpt from *Beyond Words: What Animals Think and Feel*. Copyright © 2015 by Carl Safina. Reprinted by permission of John Macrae Books, an imprint of Henry Holt and Company. All rights reserved.

Excerpt from *Silent Thunder: In the Presence of Elephants*. Copyright © 1998 by Katy Payne. Reprinted by permission of Simon & Schuster, Inc. All rights reserved.

Excerpt from "Sources of Social Complexity in the Three Elephant Species," by Katy Payne, in *Animal Social Complexity: Intelligence, Culture, and Individualized Societies*, edited by Frans B. M. de Waal and Peter L. Tyack. Copyright © 2003 by the President and Fellows of Harvard College.

Excerpt from "Self-Recognition in an Asian Elephant," by Joshua M. Plotnik, Frans B. M. de Waal, and Diana Reiss, in *Proceedings of the National Academy of Sciences of the United States*. Copyright © 2006 by the National Academy of Sciences, U.S.A.

Excerpt from "African Elephants Show High Levels of Interest in the Skulls and Ivory of Their Own Species," by Karen McComb, Lucy Baker, and Cynthia Moss, in *Biology Letters* (2006). Reprinted by permission of the Royal Society.

Excerpt from *The Elephant Whisperer: My Life with the Herd in the African Wild*, by Lawrence Anthony with Graham Spence. Copyright © 2009 by Lawrence Anthony and Graham Spence. Reprinted by permission from Thomas Dunne Books, an imprint of St. Martin's Press. All rights reserved.

Excerpt from *Elephantoms: Tracking the Elephant*, by Lyall Watson. Copyright © 2002 by Lyall Watson. Used by permission of W. W. Norton & Company, Inc.

Excerpt from "Ivory Worship," by Bryan Christy, in *National Geographic* (2012). Reprinted with permission of National Geographic Creative.

"In Praise of Pachyderms." Copyright © 2017 by The Economist Newspaper Limited, London.

Faithful Elephants: A True Story of Animals, People and War, by Yukio Tsuchiya. Text copyright © 1988 by Yukio Tsuchiya. Reprinted by permission of Houghton Mifflin Harcourt. All rights reserved.

Vu Hung, *The Story of a Mahout and His War Elephant*. Hanoi: Foreign Languages Publishing House, 1976. Reprinted courtesy of the publisher.

NOTES

CULTURAL AND CLASSICAL ELEPHANTS

1. Raman Sukumar, *The Living Elephants: Evolutionary Ecology, Behavior, and Conservation* (Oxford: Oxford University Press, 2003), 3–9.

2. Jeheskel Shoshani and Pascal Tassy, "Summary, Conclusions, and a Glimpse into the Future," in *The Proboscidea: Evolution and Palaeoecology of Elephants and Their Relatives*, ed. Jeheskel Shoshani and Pascal Tassy (Oxford: Oxford University Press, 1996), 337; Sukumar, *Living Elephants*, 3–19.

3. Colin M. Turnbull, "Legends of the BaMbuti," *Journal of the Royal Anthropological Institute of Great Britain and Ireland* 89, no. 1 (January–June 1959), 45.

4. Jan Knappert, *Myths and Legends of the Congo* (London: Heinemann Educational Books, 1971), 165.

5. Arrian's estimates of the size of the Persian army are radically higher than those of contemporary historians, per Wikipedia, "Battle of Gaugamela" entry, last edited December 17, 2019, https://en.wikipedia.org/wiki/Battle_of_Gaugamela.

6. John M. Kistler, *War Elephants* (Westport, CT: Praeger, 2006), 26–30.

7. Ibid., 34–37.

8. Ibid., 40, 41.

9. Ibid., 38–79.

10. James S. Romm, "Aristotle's Elephant and the Myth of Alexander's Scientific Patronage," *American Journal of Philology* 110, no. 4 (winter 1989), 566–75.

11. Armand Marie Leroi, *The Lagoon: How Aristotle Invented Science* (New York: Viking, 2014), 7.

12. Kistler, *War Elephants*, 80–92.

13. Ibid., 97–104.

14. Ibid., 159, 160.

15. Eric Scigliano, *Love, War, and Circuses: The Age-Old Relationship between Elephants and Humans* (Boston: Houghton Mifflin, 2002), 129.

16. Aristotle, *History of Animals*, vol. 3, edited and translated by D. M. Balme (Cambridge, MA: Harvard University Press, 1991), 215.

17. Pliny the Elder, *Natural History: A Selection* (New York: Penguin Books, 1991), 108.

18. Quoted in *The Book of Beasts: Being a Translation from a Latin Bestiary of the Twelfth Century*, ed. and trans. T. H. White (New York: Dover, 1984), 232.

19. Aristotle, *Historia Animalium*, vol. 2, ed. and trans. A. L. Peck (Cambridge, MA: Harvard University Press, 1965), 333.

20. Pliny the Elder, *Natural History*, 108.

21. Lyall Watson, *Elephantoms: Tracking the Elephant* (New York: W. W. Norton, 2002), 117.

22. Richard Conniff, "When the Music in Our Parlors Brought Death to Darkest Africa," *Audubon*, July 1987, 77–92.

23. Ibid., 83.

24. Ibid., 82.

25. Ibid., 81.

26. Ibid., 88.

27. Martin Meredith, *Elephant Destiny: Biography of an Endangered Species in Africa* (New York: Public Affairs, 2001), 106.

28. J. H. Williams, *Elephant Bill: The Best-Selling Account of 1920s Life in the Jungles of Burma* (N.p.: Long Riders' Guild Press, 2001), 26–30; Dale Peterson, *Where Have All the Animals Gone?* (Peterborough, NH: Bauhan, 2015), 87–103.

29. Charlie Campbell, "Burma's Logging Ban Is Great for Forests, but a Disaster for Its Working Elephants," *Time*, March 31, 2014, http://time.com/43317/burma-logging-ban-elephants-at-risk; "Elephants in Myanmar/Burma," EleAid, accessed December 18, 2019, www.eleaid.com/country-profiles/elephants-burma/.

30. Franklin Edgerton, trans., *The Elephant-Lore of the Hindus* (Delhi: Motilal Banarsidass, 1985), 41–48.

31. W. P. Jolly, *Jumbo* (London: Constable & Co., 1976), 28, 29.

32. Ibid., 126, 127.

33. Ibid., 135.

34. Ibid., 150, 151.

35. Shana Alexander, *The Astonishing Elephant* (New York: Random House, 2000), 120, 121.

36. Ibid., 121.

37. Ibid., 123, 124.

38. Ibid., 125.

39. Ibid., 143.

40. Ibid., 127–31.

41. Charles Edwin Price, *The Day They Hung the Elephant* (Johnson City, TN: Overmountain Press, 1992), 14.

42. Carol Bradley, *Last Chain on Billie: How One Extraordinary Elephant Escaped the Big Top* (New York: St. Martin's, 2014), 88.

43. Ibid., 72, 73.

44. Ibid., 108.

45. Richa Sharma, "India Loses One Tusker Every Four Days," *New Indian Express*, September 2, 2017.

46. Gordon G. Gallup Jr., "Chimpanzees: Self-Recognition," *Science* 167 (January 2, 1970), 86, 87.

47. Benjamin L. Hart, Lynette A. Hart, and Noa Pinter-Wollman, "Large Brains and Cognition: Where Do Elephants Fit?," *Neuroscience and Biobehavioral Reviews* 32 (2008), 86–98.

48. George Wittemyer et al., "Illegal Killing for Ivory Drives Global Decline in African Elephants," *PNAS* 111, no. 36 (2004), 13117–21; Brad Scriber, "100,000 Elephants Killed by Poachers in Just Three Years, Landmark Analysis Finds," *National Geographic*, August 18, 2014.

49. Chieko Akiyama, "To the Readers," in *Faithful Elephants: A True Story of Animals, People and War*, by Yukio Tsuchiya, trans. Tomoko Tsuchiya Dykes (Boston: Houghton Mifflin, 1988).

DALE PETERSON's twenty previous books have been named Best of the Year by the *Boston Globe*, the *Denver Post, Discover*, the *Economist*, the *Globe and Mail, Library Journal*, and the *Village Voice*. Two titles have been honored as Notable Book of the Year by the *New York Times*. He is the author of the definitive biography *Jane Goodall: The Woman Who Redefined Man*, as well as *Elephant Reflections, Giraffe Reflections, The Moral Lives of Animals*, and *The Ghosts of Gombe*. With Jane Goodall, he coauthored *Visions of Caliban: On Chimpanzees and People*, and he is coeditor, with Marc Bekoff, of *The Jane Effect: Celebrating Jane Goodall*, published by Trinity University Press. A former fellow at the Radcliffe Institute for Advanced Study at Harvard University, Peterson teaches at Tufts University and lives in Arlington, Massachusetts.

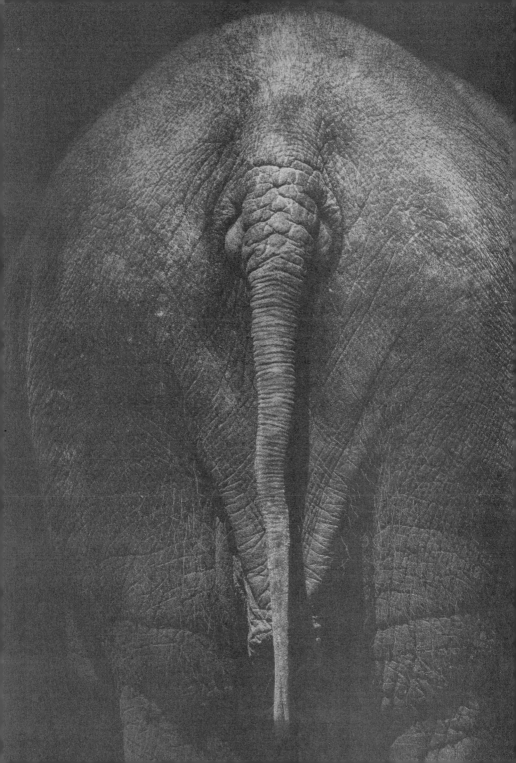